Advance Directives

INTERNATIONAL LIBRARY OF ETHICS, LAW, AND THE NEW MEDICINE

Founding Editors

DAVID C. THOMASMA†
DAVID N. WEISSTUB, *Université de Montréal, Canada*
THOMASINE KIMBROUGH KUSHNER, *University of California, Berkeley, U.S.A.*

Editor

DAVID N. WEISSTUB, *Université de Montréal, Canada*

Editorial Board

VOLUME 54

For further volumes:
http://www.springer.com/series/6224

Peter Lack • Nikola Biller-Andorno
Susanne Brauer

Editors

Advance Directives

Springer

Editors
Peter Lack
Department of Moral Theology
 and Ethics
University of Fribourg
Basel, Switzerland

Nikola Biller-Andorno
Institute of Biomedical Ethics
University of Zurich
Zurich, Switzerland

Susanne Brauer
Institute of Biomedical Ethics
University of Zurich
Zurich, Switzerland

ISSN 1567-8008
ISBN 978-94-007-7376-9   ISBN 978-94-007-7377-6 (eBook)
DOI 10.1007/978-94-007-7377-6
Springer Dordrecht Heidelberg New York London

Library of Congress Control Number: 2013951646

Printed on acid-free paper

Springer is part of Springer Science+Business Media (www.springer.com)

Preface

Advance directives have been intensely debated since the 1980s. Over the past few years, a growing number of articles have focused not only on general ethical questions but also on matters of implementation, as advance directives have become increasingly widespread in institutions such as hospitals and nursing homes. The particular issues arising depend on various contextual factors, such as the legal framework or the attitudes of health professionals regarding patient autonomy and its limits.

One of the main reasons for the rise of advance directives has been progress in medical treatments and new technologies allowing patients to survive in life-threatening situations that would previously have led to death. Although these advances are considered beneficial in many cases, some patients want to exclude treatments which could be futile and stand in the way of a peaceful death. Moreover, the value attached to patient autonomy—and the rejection of medical paternalism—has been growing in recent decades. Respect for patient autonomy, which is crucial in current medical ethics and legislation, also embraces the right to refuse treatment. If patients are unable to express their wishes because of mental incapacity, advance directives are frequently used in surrogate decision-making.

Given the broad fundamental consensus on the appropriateness of advance directives, one might wonder if there was anything left to clarify. However, a closer look at the debate on advance directives reveals that many ethical and policy issues remain in need of further exploration.

One concern is the ethical legitimation of advance directives. Their authority, in particular, is still widely debated, both from an ethical and more recently also from a legal viewpoint. This point is well illustrated by the example of Austrian legislation, which distinguishes between advance directives which are binding (*verbindliche Patientenverfügung*) and those which are (merely) to be taken into consideration by physicians (*beachtliche Patientenverfügung*). With regard to the German legislation, it was widely discussed whether it should be possible for advance directives to be drawn up for any (medical) situation, or whether some conditions, such as persistent vegetative state, should be excluded from this kind of patient self-determination. The issues of consequences for family members and possible conflicts with medical and nursing ethical standards are also subject to ongoing debate. As long as doubts concerning their ethical legitimation persist, advance directives will continue to be questioned—socially, medically and politically.

Opinions differ widely on how advance directives should be legally implemented; for example, there are major differences between countries concerning the (legally) binding force of advance directives, or the particular situations for which they can be established. How decision-making capacity is to be determined and/or documented when advance directives are prepared is also a matter of debate. In addition, various solutions have been proposed for specific problems associated with advance directives in the context of dementia care, imprisonment and psychiatry.

Attitudes towards advance directives differ not only between countries but also within a given culture or state: groups such as patients or healthcare professionals may tend to adopt different perspectives, and matters become even more complex when individuals from a different cultural background are confronted with biomedical or end-of-life questions.

This volume presents an overview of the issues arising with regard to advance directives from an international perspective. Here, the familiar and widely accepted concept of patient self-determination is understood in a broader sense to cover future situations. This includes instruments such as the power of attorney for healthcare (i.e. an individual designated by the patient) and advance care planning, which encompasses both living wills and medical/nursing care aimed at preserving patients' autonomy in a future situation of mental incapacity due to the progression of a given illness. It is hoped that the identification and discussion of common themes and differences in the understanding of advance directives, as well as regulations, policies and clinical practices, will contribute to a continuous process of improving patient care while also promoting respect for patients' own preferences.

In Part I, we focus on the development of advance directives and consider the prerequisites for their validity. In order to comprehend the ethical issues involved and the ongoing controversies, it is essential first to review the history of the concept.

Alfred Simon describes the emergence of advance directives as a means to secure patient autonomy even in situations where patients are no longer competent or able to express their preferences. This historical account starts with the origins of advance directives in the USA and goes on to describe their adoption in various European countries. It is emphasized that the implementation and legal status of advance directives is heterogeneous, despite the common normative framework provided by the Council of Europe's Oviedo Convention. Simon's chapter concludes with an overview of the main arguments brought forward in favour of and against advance directives and legal regulation in this area.

Also considered in Part I are issues relating to the validity of advance directives—the prerequisites that are philosophically disputable and contested in practice under different legal regulations. Decision-making capacity is a crucial concept for advance directives: individuals need to be competent when they write an advance directive which will only take effect when they have lost their decision-making capacity. The assessment of patients with mental disorders is particularly complex.

Marie-Jo Thiel investigates whether individuals are indeed able to anticipate preferences concerning future illness. Anticipation, she argues, is influenced by social and cultural factors. A brief historical analysis highlights the wish to remain in control, even at the time of one's own death, and the reliance on reason as major elements underlying the current interest in advance directives. Further contextual factors are identified in situational and philosophical analyses. Finally, the difficulties of anticipating future preferences are illustrated with reference to two examples—amyotrophic lateral sclerosis and Alzheimer's disease.

Jochen Vollmann presents and discusses different concepts of competence, assessment tools and associated challenges in the light of relevant empirical studies. He concludes with ethical recommendations for the use of advance directives in psychiatric practice, emphasizing the potential of advance directives to improve patient-oriented psychiatric care by fostering communication between patients and mental health professionals.

Apart from the fundamental conceptual issues, a number of questions arise concerning the implementation of advance directives. For example, should there be any limit to the treatment options that can be refused or specified in advance directives? How binding should they be—do physicians have to consider them or act upon them (unless certain exceptions apply)? Should incompetent persons' non-verbal gestures and signs render advance directives invalid? Are there situations where advance directives could be disregarded because of certain constraints (e.g. prisons, psychiatric interventions). These issues are discussed in Part II.

Robert Olick first examines the scope and limits of advance directives, together with the rights and duties of healthcare providers and healthcare proxies. A number of cases are used to illustrate common challenges for honouring advance directives in different clinical contexts, and to demonstrate how the ethical and legal perspectives may collide.

Whereas the criteria for the validity and binding force of advance directives have been increasingly regulated, the conditions for revoking an advance directive are less clear. Ralf Jox addresses this issue, first discussing the stability of treatment preferences over time and then examining possible conditions for valid revocation. Finally, he considers the question of whether non-verbal, behavioural expressions can be interpreted as constituting a revocation of an advance directive.

Advance directives and advance planning play a special role in psychiatry. The fluctuating nature of mental illness gives rise to difficulties in the writing and implementation of advance directives. According to Jacqueline Atkinson and Jacquie Reilly, advance directives in psychiatry will always involve a conflict between self-determination (i.e. following the patient's wishes) and society's desire to protect patients and their fellow citizens. This conflict is exacerbated when compulsory detention and treatment are required but have to be balanced with the patient's right to autonomy.

Bernice Elger's contribution deals with a particular patient population— imprisoned patients. The widely acknowledged principle of equivalence stipulates that prisoners are to receive the same standard of healthcare as is available to the general population. This includes the right to self-determination regarding medical interventions, which may take the form of an advance directive. Elger first examines general ethical and legal aspects of advance directives in the context of imprisonment and then discusses advance directives in three specific situations—hunger strikes, end-of-life care and psychiatric treatment of detained persons.

It is not clear precisely how advance directives affect the relationship between patient and physician, family members and loved ones. Some healthcare workers consider their professional ethos and autonomy to be called into question by advance directives, and relatives or loved ones may be surprised or even hurt by the wishes specified in advance directives. Yet some patients assert that their trust in a physician is strengthened by discussing and establishing advance directives, and that, by giving instructions in this way, they intend to ease the burden of decision-making otherwise borne by their family and loved ones. Part III broadens the perspective to include the social environment of those who use advance directives.

Mark P. Aulisio argues that the advance directive movement, while failing to achieve its primary goal, has nevertheless brought about a paradigm shift in the patient-physician relationship. In what he terms the "standard justification", advance directives are taken to be vehicles for making effective the autonomous *moral* agency of persons in circumstances in which such agency could not otherwise be effective since the person in question lacks the capacity to exercise this agency. In empirical terms, the advance directive movement has clearly missed its goal since only a minority of patients possess an advance directive. Also, surveys have demonstrated that patients' treatment preferences are often incorrectly predicted by surrogate decision-makers. However, by supporting patient autonomy as a *political* value—a limitation of the authority and standing of the "good doctor"—the advance directive movement has promoted a more patient-centred approach in clinical practice.

Focusing on the role of the family and close persons, Margot Michel considers the international legal framework and Swiss legislation on advance directives. How, she asks, is patient autonomy to be protected when a patient becomes incompetent? Under these circumstances, the role of the family and close persons becomes crucial because advance directives need to be interpreted and implemented by third parties. Michel stresses that the decision to declare a patient incompetent is absolutely vital to the further involvement of the patient in the decision-making process. She recommends that, for the assessment of competence, ethical guidelines and best practice standards should be developed, as opposed to a rather inflexible legal definition. However, the law can support patient autonomy by specifying who can act as a proxy decision-maker for an incompetent patient, and legal safeguards are required in case the patient's representative or medical staff fail to act in the patient's interests or according to his or her presumed wishes.

Settimio Monteverde analyses advance directives in the context of nursing care. He points out that nurses have a privileged position deriving from their proximity to the patient. This position can lead to ambiguities when instructions given in an advance directive conflict with medical prescriptions, proxies' preferences or the ethos of good nursing care. Nurses are called on to express their ethical concerns when care-related conflicts arise. In addition, they need to resolve the tension between closeness and distance by respecting the patient's wishes and best interests without failing to question an advance directive if necessary. Because of their special closeness to patients, nurses can act as systemic change agents in establishing and maintaining a culture of trust and patient orientation at every stage of the advance planning and delivery of care.

Part IV takes up the question of ethical challenges raised by advance directives. According to Manuel Trachsel, Christine Mitchell and Nikola Biller-Andorno, the implementation of advance directives depends on their interpretation by third parties, namely the responsible physician or healthcare team. Advance directives are not always clearly formulated and have to be interpreted in the light of the specific medical situation. This process may involve a conflict between respect for autonomy on the one hand and paternalism on the other. As well as presenting legal standards, the authors discuss various ethical criteria that can provide guidance for appropriate interpretation of advance directives.

Ruth Horn and Ruud ter Meulen tackle the question of whether it is ethically justified to promote advance directives as instruments for cutting the costs of healthcare by counteracting the growing medicalization of dying. The authors emphasize that the use of advance directives for cost control is only ethically acceptable if they reflect patients' authentic wishes. In their view, however, it remains questionable whether advance directives are valid instruments to express the patient's genuine will in a specific situation. In order to avoid shortcomings of advance directives such as the difficulty of anticipating future events and specific preferences, Horn and ter Meulen recommend that advance directives should be placed in the broader context of advance care planning, aiming to enhance conversations between physicians and patients.

Some of the authors in this volume see the term "advance directive" also as a code for a cultural shift in medicine, away from paternalism and the primacy of the technologically feasible towards a practice centred on the patient's needs and preferences. Tanja Krones and Sohaila Bastami outline the potential for advance directives to evolve from a legal—sometimes legalistic—document with limited utility in clinical practice into a patient-oriented process emphasizing communication and support for patients and caregivers. Starting with a discussion of the reasons for the "failure" of the living will, their chapter describes the main features of advance care planning, obstacles to its implementation and empirical data on its effectiveness.

In the concluding remarks, the editors raise the question of whether it is culturally and politically desirable and ethically required to try and reach a more substantial agreement on advance directives beyond the minimal consensus formulated in the 1997 Convention on Human Rights and Biomedicine. They identify common ground between the individual contributors and highlight the importance of a relational understanding of advance directives, as well as the broader clinical context of advance care planning, rather than a narrow, legalistic approach. Finally, they call for specific legal safeguards and for common standards to determine when an advance directive goes into effect.

Much of the important literature on advance directives was published in the 1990s. These publications no longer represent the current state of the debate, either in bioethics or with regard to recent European developments in policy and law on advance directives. The present volume—which had its origins in the international conference entitled *Advance Directives: Towards a Coordinated European Perspective?* hosted by the Institute of Biomedical Ethics at the University of Zurich in 2008—is designed to fill a gap by providing a synopsis of the current major ethical and policy issues in a systematic manner and from a multidisciplinary perspective. We hope it will be a useful resource for those studying or dealing in practice with the many facets of advance directives.

Finally, we wish to thank the European Science Foundation, the University of Zurich's University Research Priority Program for Ethics and the Swiss National Science Foundation for their generous support.

February 2013   Nikola Biller-Andorno
                                                                                      Susanne Brauer
                                                                                         Peter Lack

Contents

Part I
History of Advance Directives and Prerequisites for Validity

Chapter 1
Historical Review of Advance Directives

Alfred Simon

1.1 Introduction

The discovery of antibiotics and chemotherapy, as well as the development of medical techniques and interventions such as mechanical ventilation, artificial nutrition and hydration or organ transplantation, led to an enormous increase in the capacity of medicine to sustain human life even under the most difficult conditions. But these advances are not always beneficial for the individual patient; for some, they lead to a prolongation of suffering and dying. Physicians have to perform a balancing act between what is medically and technically possible and what is humane and ethically acceptable.

Until a few decades ago, physicians involved in ethical decision-making primarily considered the principles of beneficence and non-maleficence. Respect for the patient's autonomy clearly played a subordinate role. In traditional ethical codes such as the Hippocratic Oath or the Declaration of Geneva of the World Medical Association, no reference is made to the patient's wishes. This has changed radically: in line with the shift in values towards greater individuality and personal responsibility observed in the Western world in the second half of the twentieth century, patient self-determination became more and more important. Legal decisions and modern medical ethics emphasize that patient autonomy overrides what physicians consider best for the well-being of the patient. Informed consent is now a widely acknowledged normative standard in medical ethics (Faden and Beauchamp 1986).

Advance directives are expressions and consequences of these developments. They allow patients to exercise their right of self-determination even in situations where they are no longer able to communicate or decide. The idea of advance

A. Simon (✉)
Academy for Ethics in Medicine, Göttingen, Germany
e-mail: asimon1@gwdg.de

P. Lack et al. (eds.), *Advance Directives*, International Library of Ethics,
Law, and the New Medicine 54, DOI 10.1007/978-94-007-7377-6_1,
© Springer Science+Business Media Dordrecht 2014

directives arose in the US in the late 1960s and became widespread in the following years. Today, advance directives are established not only in every state of the US but also in many European countries. Nevertheless, they remain controversial.

1.2 Advance Directives in the US

This historical review of advance directives starts with an account of the origins and development of these documents in the US.

1.2.1 Origins of Different Kinds of Advance Directives

The original form of an advance directive is the so-called living will. This term was proposed by the Illinois human rights lawyer Luis Kutner. In a paper published in 1969, he suggested that, just as the last will and testament contains instructions to be followed in the event of a person's death, a person's preferences concerning medical treatment could be recorded in case of loss of capacity to consent (Kutner 1969). In certain situations, this document—referred to as a living will—was to release the physician from the duty to preserve life. Kutner recommended that the document should be notarized and attested to by at least two witnesses, affirming that the maker was of sound mind and acted of his own free will.

The Euthanasia Society of America took up Kutner's suggestions and printed the first version of a living will form in 1972. The form was a short and relatively simple text, stating that, if there is no reasonable expectation of recovery from physical or mental disability, the signatory wishes to be allowed to die and not to be kept alive "by artificial means or heroic measures". Furthermore, the form included a request that drugs be "mercifully administered ... for terminal suffering even if they hasten the moment of death" (Zucker 1999). Having no legal force at that time, the request was rather an appeal to the moral sensibilities of the physician or other people caring for the patient.

In 1976, the Natural Death Act was passed in California. This was the first law dealing with the legal authority of living wills. In addition, it strengthened the right of terminally ill patients to refuse life-sustaining measures. The law was occasioned by the public controversy concerning the case of Karen Ann Quinlan. In a landmark decision in the case of this young, comatose patient, the New Jersey Supreme Court ruled that the constitutionally protected right of privacy outweighs the interest of the state in preserving life. As her legal guardian, Karen's father was thus considered eligible to exercise this right on her behalf. Other states followed the example of the Californian legislation in the late 1970s and 1980s. Today, corresponding laws are to be found in every state of the US. Many of these laws also include drafts for the formulation of living wills, which are mainly couched in very general language.

The non-specific wording of many living will forms, and the resulting difficulties in following directives in an acute treatment situation, led to the development of a second type of advance directive—the durable power of attorney for healthcare. In this document, individuals can appoint a trustworthy person as their healthcare proxy, someone who can communicate with the physician and decide on treatment measures when the patient is no longer able to do so. When the patient has written a living will, the healthcare proxy's main task is to ensure that physicians act in accordance with it, and possibly to assist in interpreting it with regard to the specific situation. The durable power of attorney for healthcare should therefore be regarded as an addition, rather than as an alternative, to a living will. The first state to have the possibility of appointing a healthcare proxy enshrined in law was Pennsylvania in 1983. Today, similar regulations are in force in almost every state (Brown 2003).

Most US states also have surrogate decision-making laws, defining who is eligible to act as a proxy for a patient who is unable to consent and has not prepared an advance directive. In general, this will be the spouse, adult children, adult siblings, other close relatives or other persons close to the patient (Meran et al. 2002). It is interesting to note that, under the 1999 Texas Futile Care Law, a healthcare facility is allowed to discontinue life-sustaining treatment against the wishes of the patient or healthcare proxy 10 days after giving written notice, if continuation of treatment is considered medically futile by the responsible medical team (Lehmann 2008).

1.2.2   Promotion and Distribution of Advance Directives

Surveys in the late 1980s showed that few people were informed about the possibility of an advance directive and even fewer had completed one (Damato 1993; Cugliari et al. 1995). At the same time, the US Supreme Court ruling in the case of Nancy Cruzan highlighted the importance of previously expressed or presumed wishes when a person is not currently able to give consent in the context of decisions on maintaining or ending life-sustaining treatment (Zucker 1999). Against this background, the Patient Self-Determination Act was passed by the US Congress in 1990 in order to promote public knowledge and awareness of advance directives. This law is binding for all hospitals, hospices and nursing homes funded by the national Medicaid or Medicare programmes. Provision is made for institutions to inform their patients or residents in writing on their rights concerning end-of-life decision-making and advance directives. The existence of a patient's advance directive has to be documented in the patient's medical records. Furthermore, institutions are obliged to provide for education for staff and the community on this issue.

Several studies demonstrated that better education on advance directives, e.g. aided by written information and/or personal counselling, can help to increase the number of these documents completed in practice (Rubin et al. 1994; Meier et al. 1996). The key question is whether this development leads to better quality in medical care of terminally ill patients. The findings of the large-scale SUPPORT study (Study to Understand Prognoses and Preferences for Outcomes and Risks of Treatment; Teno

et al. 1997) were rather sobering: though provision of information increased documentation of existing advance directives, it did not substantially enhance the quality
of physician-patient communication or medical decision-making. The authors
concluded that merely increasing the frequency of advance directives is not an
effective way of improving the medical care given to the seriously ill. Instead, they
advocated better communication and more comprehensive advance care planning.

Despite all the national and institutional efforts to inform the public about
advance directives, their distribution is still rather limited in the US. Several studies
show that between 18 and 30 % of US citizens have prepared a living will or a
durable power of attorney for healthcare. The average author of an advance
directive is white, with higher socioeconomic status and a higher burden of illness,
is well informed about the possibility of advance directives and usually has a long-
time relationship with his or her primary care physician (Wilkinson et al. 2007).

1.3 Advance Directives in Europe

In comparison to the US, the European debate on advance directives began with
a few years' delay. The significance of advance directives still differs regionally: in
some countries, legal provisions regulate the authority of living wills and the
appointment of healthcare proxies, while in other countries advance directives are
irrelevant in medical practice and are not even subject to debate in medical ethics or
medical law (Brauer et al. 2008).

In 1997, the European Convention on Human Rights and Biomedicine, also
known as the Oviedo Convention, was adopted by the Council of Europe (1997a).
It is the first legal instrument to introduce international—albeit minimal—regulation
of advance directives. According to Article 9 of the Convention, the previously
expressed wishes of a patient not able to express his or her wishes at the time of
a medical intervention are to be "taken into account". However, this does not mean
that these wishes should necessarily be followed. The Explanatory Report states
that there may be grounds for not heeding the patient's opinion "when the wishes
were expressed a long time before the intervention and science has since progressed"
(Council of Europe 1997b). Nevertheless, this provision at least obliges physicians
or relatives to justify any treatment decision running counter to the patient's wishes.
To date, the Oviedo Convention has been signed by 35 of the 47 member states of
the Council of Europe, and ratified by 29.

1.3.1 Overview of the Legal Situation in Europe

Table 1.1 provides an overview of the current legal situation concerning advance
directives in various European countries. What is striking is that national support

Table 1.1 Overview of the legal situation concerning advance directives in Europe

Country	Legal regulations	Oviedo convention ratified (signed)	Legal force of living wills	Option of appointing a healthcare proxy
Austria	Living Will Act (2006)		Yes[a]	Yes
Belgium	Act on Patients' Rights; Act on Euthanasia (2002)		Yes	Yes
Bulgaria		2003	No	No
England and Wales	Mental Capacity Act (2005)		Yes[a]	Yes
Finland	Act on the Status and Rights of Patients (1992)	2009	Yes	No formal regulation
France	Law n°2005-370 (2005)	2011	No	Yes[b]
Germany	Amendment of the Guardianship Law (2009)		Yes	Yes
Greece		1998	No	No
Hungary	Health Care Act (1997)	2002	Yes[a]	Yes
Italy	Law on Living Wills (passed in 2011)	(1997)	No	No
Lithuania		2002	No	No
Netherlands	Medical Treatment Contract Act (1995); Dutch Euthanasia Act (2002)	(1997)	Yes	Yes
Norway		2006	No	Yes[b]
Portugal		2001	No	No
Serbia		2011	No	Yes
Slovakia		1998	No	No
Spain	Regional laws (since 2000)	1999	Yes[a]	Yes
Switzerland	Law on the Protection of Adults (2013)	2008	Yes	Yes
Turkey		2004	No	No

[a]To be legally binding, strict formal and procedural conditions have to be met
[b]Proxy has no authority to refuse medical treatment

for the Oviedo Convention does not seem to have a direct influence on the legal force of living wills. In some signatory states (e.g. Bulgaria, France, Greece, Italy, Lithuania, Norway, Portugal, Serbia, Slovakia, Turkey), living wills are not legally binding, while in other countries—that have not signed the Convention—living wills are highly binding (e.g. Belgium, Germany). There are also considerable differences among the applicable legal regulations: in Belgium, Finland, Germany, the Netherlands and Switzerland, living wills are highly binding; in Austria and Hungary, living wills are binding only under strict formal and procedural conditions; and in France living wills are not legally binding, but merely advisory. Legal provision for the appointment of a healthcare proxy is only made in a few European countries.

1.3.2   Examples from Various Countries

In this section, the variety of legal regulations in Europe is illustrated by a number of examples.

1.3.2.1   Italy

The Italian Parliament passed a law on living wills in July 2011, after more than 2 years of contentious debate on this regulation. The law prohibits any kind of euthanasia and contains various restrictions concerning the legal force of living wills. Living wills will be applicable only when the patient is comatose. Otherwise, a written living will is not legally binding for the physician. It is prohibited to withhold artificial nutrition or hydration, even if this is expressly requested by the patient. Life-prolonging treatments have to be commensurate with the therapeutic aims. The debate on this law was strongly influenced by the case of Eluana Englaro, a 38-year-old patient who was in a coma for 17 years and died in February 2009 after her father obtained the consent of the Supreme Court of Cassation for artificial nutrition to be terminated (Turone 2009, 2011).

1.3.2.2   France

Living wills are not legally binding in France either. Under the 2005 law, end-of-life decisions concerning persons incapable of consent must be made by physicians. Living wills in which the patient has expressed his or her wishes concerning such treatment decisions are only one element among others that have to be taken into account. This means that advance directives are merely advisory, not binding. In addition to the living will, which has to be renewed every 3 years, the patient may appoint a trusted person as healthcare proxy. The healthcare proxy has to be consulted by physicians but has no authority to refuse medical treatment (Binet 2008).

1.3.2.3   Austria

The Austrian Living Will Act, which came into force in 2006, makes a distinction between binding and non-binding living wills. For the creation of a binding living will, strict formal and procedural conditions have to be met: the living will has to be drafted in writing and dated in the presence of a notary or representative of a patient advocacy organization. The medical treatments which are refused must be described in detail or must be clearly apparent from the overall context of the document. Comprehensive medical advice is a prerequisite for a living will. The physician has to document the consultation and confirm the patient's capacity for insight and judgement. The physician also has to state the fact that and the reasons why the patient appropriately evaluates the consequences of his/her living will. The living will must be renewed after 5 years at the latest. Living wills that do not meet these conditions are not directly legally binding for the physician, but they have to be taken into account when determining the patient's presumed wishes— the more they meet the conditions of a binding living will, the more they have to be taken into account. That means that in practice they may have the same force as a formally binding living will. Binding and non-binding living wills are invalid if they are based on error, fraud or deception, or on physical or mental constraint; if the content is legally unacceptable; if there has been a substantial change in the state of medical science with regard to the content of the living will; or if the living will is revoked by the patient him/herself (Aigner 2007). Patients may appoint a healthcare proxy. Close relatives are permitted to consent to minor interventions that do not lead to serious or lasting detriment to the physical integrity or personality of the patient, if the patient is not able to give consent (Inthorn and Kletecka-Pulker 2008).

1.3.2.4   Germany

In Germany, the legal force of living wills is regulated by the 2009 Amendment of the Guardianship Law. Living wills, understood as previous written expressions of treatment preferences by a person with the capacity to consent, are legally binding regardless of the nature and stage of disease. If a patient is unable to give consent, the healthcare proxy, appointed by the patient or the guardianship court, has to examine whether the living will is consistent with the patient's current life and treatment situation. If this is the case, the healthcare proxy has to ensure that the patient's wishes are understood and complied with. Revocation of a living will is possible at any time, without any formal requirements. If no written living will is available or if the living will does not match the current situation, the healthcare proxy has to decide on the basis of the patient's presumed wishes. These are to be ascertained on the basis of concrete indications, especially verbal or written statements, ethical or religious beliefs and other personal values of the patient. In the determination of the patient's presumed wishes, discussion between the healthcare proxy and the physician is mandatory. Relatives and other trusted persons are to be given the opportunity to express their views in this discussion. The appointment of a healthcare proxy was already possible before the latest amendment. If no proxy is named by the patient, such a person has to be appointed by the court. Unless they are appointed by the patient or the court, close relatives cannot act as a proxy for adult patients (Simon 2010).

1.3.2.5   Switzerland

In Switzerland, a law on the protection of adults came into force on 1 January 2013. This represents the country's first uniform (federal) regulation of the legal force of living wills. Previously, various cantonal regulations coexisted, some of them differing considerably (Brauer 2007). The new law aims to strengthen the citizen's self-determination by the use of instruments such as living wills. A person's living will is now legally binding for the physician, as long as it is based on the patient's free will and does not call for prohibited measures (e.g. euthanasia), and as long as there is no indication that the patient's wishes may have changed (Geth 2010). The new law makes provision for the designation of a healthcare proxy and for allowing close relatives to make decisions on behalf of the patient. Previously, in most cantons, only physicians had the legal authority for surrogate decision-making.

1.3.2.6 Spain

In Spain, advance directives are regulated by a national law dating from 2002 and by further regional laws. Under the national law, the author of a living will is required to be both competent and over 18 years of age. In some regions, even minors from the age of 16 may create a directive. The living will may contain statements about the type and extent of medical treatment desired, as well as covering organ and tissue donation. Insofar as living wills refer to medical treatment and care, they are subject to certain restrictions: They must not be implemented if they conflict with the legal system or with good clinical practice, or if the circumstances do not match the clinical event or situation envisaged by the patient. Furthermore, the legal force of living wills is subject to certain procedural conditions: in some regions, for instance, they have to be officially registered. It is possible to appoint a healthcare proxy, even though his/her authority to decide is not conclusively regulated and is a matter of controversy (Seoane and Simón 2008). One of the notable features of the 2002 law is that it provides for the creation of a national register for advance directives, linking the two existing regional registers.

1.3.2.7 The Netherlands

What is unique about the regulations in the Netherlands is that, as well as refusing life-sustaining treatment in a living will, patients may even state a desire for euthanasia. The possibility of refusing medical treatment in advance directives has existed since 1995, when some provisions of the Civil Code were modified by the Medical Treatment Contract Act. No special legal conditions have to be met for a living will to be binding. It is stated in the law that doctors are not obliged to comply with directives if they have "well-founded reasons"; however, it is not specified what sort of reasons may justify non-compliance. In clinical practice, this leads to a tendency to disregard especially the wishes of demented patients requesting limitation of treatment (Rurub 2008). The Medical Treatment Contract Act provides for the appointment of a healthcare proxy. If no proxy is named by the patient, the patient's spouse or partner or another close relative has to act as a proxy. In 2002, the Law on Euthanasia and Assisted Suicide came into effect. This allows physicians to help patients to actively end their life under certain specified conditions. Patients can record their wishes in a euthanasia-specific advance directive in case they lose their decision-making ability. However, such directives are not legally binding for physicians.

1.4 The Debate on Advance Directives

Having illustrated the situation in various countries, we now consider a number of arguments typically advanced for or against advance directives and legal regulation in this area.

1.4.1 Arguments in Favour of Legal Regulation of Advance Directives

What is considered the most important argument for legal regulation of advance directives is the establishment and protection of the patient's right of self-determination. The argument starts from the premise that not only can the patient determine medical treatment here and now, but the right to self-determination extends to future situations when the patient is unconscious and incompetent. The patient's consent given prior to a surgical operation performed under general anaesthesia refers to the same premise, but is usually not called into question. What is regarded as valid when consenting to medical interventions—so the proponents' argument runs—must also apply to the refusal of medical treatment.

Another argument in favour of legal regulation of advance directives is the increase in legal certainty for all concerned. Patients want to feel certain that their wishes will be respected, and physicians wish to be secure in the knowledge that their compliance with the patient's wishes will not have adverse consequences for them. Both are dependent on legal regulation of the authority of, and conditions for, advance directives.

1.4.2 Fundamental Objections to Advance Directives

Critics of advance directives, for their part, do not consider it possible or even desirable to determine binding directives for future and unforeseeable situations. They point to the fact that medical conditions such as advanced dementia or a vegetative state are associated with such a profound personality change that mandatory compliance with advance directives amounts to enslavement of the current person by the former person. This statement is justified by invoking commonplace experiences that find expression in phrases such as "He is no longer the same person he used to be before his illness", but also with reference to theoretical deliberations on the identity of persons. The philosopher Derek Parfit (1984) stated that personal identity is less a question of continuity over time and place than a matter of continuity of our memories, intentions and wishes. When the degree of psychological continuity falls below a certain minimum level, the person's identity is disrupted, and we can no longer speak of "the same person". Applied to advance directives, this would mean that a directive created in healthy times would, in the event of a change of personality due to illness, only be applicable to another person.

Parfit's thesis was revisited and modified by Allen Buchanan and Dan Brock. They, too, regard psychological continuity as a necessary condition for personal identity. However, unlike Parfit, who assumes a gradual change of identity, Buchanan and Brock (1989) postulate that a discontinuity of personal identity only occurs when neurological damage is so severe that one cannot even speak of the patient as a person. As long as this threshold is not crossed, advance directives may be considered to represent the interests of the very same person. And even if the threshold is passed, advance directives retain a certain authority because of the intimate relationship obtaining between what are acknowledged to be distinct individuals.

Although these theses on personality are highly controversial in the medico-ethical debate and do not reflect the author's own views, they call attention to important problems associated with advance directives.

1.4.3 Practical Problems Concerning Advance Directives

Apart from the fundamental objections, various practical problems may arise concerning the creation of and adherence to advance directives; these are rated differently by supporters and opponents with regard to the possibility of creating binding treatment directives for the future (Meran et al. 2002; Simon 2010).

One problem in creating advance directives is the unpredictability of future situations. What diseases will occur? What will be the specific diagnosis and prognosis? What treatment options will be available, involving what chances and risks? These are questions of high importance for future decision-making, but they cannot be (readily) foreseen at the time when an advance directive is created. This is especially true of advance directives created in healthy times. Persons with advanced disease, however, are often able to clearly foresee the future situations to which their advance directives apply and are therefore able to draft the directive with special regard to those situations. The problem of a lack of predictability does not apply exclusively to advance directives, but arises in many other situations in human life (e.g. education, career choice, starting a family). Another problem is that of limited dialogical decision-making. While a competent patient is able to make difficult decisions together with the attending physician and can reconsider and revise these decisions at any time if necessary, the incompetent patient does not have these options. Statements in advance directives based on an irrational fear of treatment, or on overestimation of treatment options, cannot be revised in later situations. On the other hand, an advance directive can build a bridge between the physician and patient in a situation lacking ways of communication: if the physician did not know the patient well before, the advance directive will provide information on the latter's ideals, wishes and fears.

As regards adherence to advance directives, the problem lies in the fact that they often include statements that are ambiguous and therefore require interpretation. This problem stems from a general dilemma that every author of an advance directive has to face: if one opts for general phrasing, the advance directive cannot serve as a binding instruction, and one's wishes may not be observed or may be misinterpreted. On the other hand, if one creates a detailed list of very specific situations and treatment preferences, there is a risk that a different situation will occur—which was not foreseen and therefore not decided upon in advance. Against this background, many legal regulations allow the patient to appoint a proxy, who can help to interpret general statements in relation to a specific treatment situation. This requires the proxy to be familiar with the patient's wishes and preferences, which is—as shown by empirical studies—not always the case (Fagerlin et al. 2001).

Another objection raised against advance directives is that values, attitudes and decisions may vary in different stages of life. Physicians and carers often observe that the patient's evaluation of conditions and impairments caused by illness changes over time, depending on whether the patient is healthy or in later stages of disease. But even patients with advanced disease may change their views regarding certain treatments, depending on their current physical constitution and psychological condition, the course of disease or changes in their social environment. Then again, the creation of an advance directive involves an examination and clarification of one's own treatment preferences, so that patients who have created an advance directive may be considered to have more stable value judgements.

A particular problem arises when present expressions of the will to live seem to contradict formerly expressed wishes, as in the case of a dementia patient obviously enjoying life but who has refused life-sustaining measures in an advance directive: in critical situations, should the current "natural" will or the instructions given in advance be observed? While some allow bodily expressions (as manifestations of the will to live) to override previously given instructions, others insist on the need to adhere to the earlier, conscious expressions of the patient's wishes—especially when the advance directive explicitly states that bodily expressions of the will to live are not to be considered decisive.

1.4.4 Autonomy and (Medical) Paternalism

The problems mentioned above clearly need to be taken seriously. There are no easy solutions; rather, they call for a weighing of interests on the part of society. How much importance should be given to the patient's right to self-determination, how much to the duty of physicians and the state to promote the patient's well-being and to protect human life? In other words, what is the appropriate balance between autonomy and care? It is generally true that patients exercising and implementing their autonomy are dependent on the caring support of other people. The appropriate balance between autonomy and care would thus seem to be unproblematic, as long as care aims at the empowerment, maintenance or re-establishment of autonomy. The crucial question is to what extent medical and state interventions may be justified under the label of care. This question is the subject of the medico-ethical debate on paternalism (Childress 1982; Schöne-Seifert 1996).

The term "paternalism" encompasses measures aiming at the protection of the alleged well-being of persons even against their actual wishes and preferences. The ethical debate distinguishes between different kinds of paternalistic action. According to the opponents of such action, a distinction can be drawn between "strong" and "weak" paternalism. If the persons concerned have the capacity to decide autonomously on their affairs (i.e. are able to give consent), it is a case of strong paternalism; if those concerned are not able to give consent, it is a case of weak paternalism (Feinberg 1971). Focusing on the paternalistic measures envisaged, a second distinction can be made between "hard" and "soft" paternalism. Hard paternalism covers measures such as legal prohibitions or physical force whereby someone is prevented from doing something considered harmful for them. Soft paternalism refers to measures designed to indirectly influence a person's behaviour and preferences, e.g. by providing information on risks or increasing the costs of an activity (Patzig 1989).

If these distinctions are applied to the various advance directive regulations, it can be seen that they represent different forms of paternalism. The restriction that only persons able to consent are allowed to create an advance directive can be regarded as a weak form of paternalism. Its ethical justification is that it does not constrain the rights of persons able to consent, but still protects people from the consequences of non-autonomous decisions.

The same is true for the provision that physicians do not have to comply with a patient's former wishes if there is good reason to believe that the patient would not stand by his or her directive owing to a change of circumstances (e.g. new treatment options). This provision is widely regarded as justified, especially when combined with certain conditions (e.g. consensual decision-making with the patient's healthcare proxy) that prevent a tendency towards general disregard of living wills (invoking the mere possibility that patients could change their mind).

Procedural requirements for the validity of advance directives (such as medical consultation, periodic renewal, the presence of a notary, etc.) are manifestations of strong paternalism in a soft form. They impose constraints on people exercising their right to self-determination so as to ensure that the documents contain the current wishes of the informed and competent patient. Discussion concerns their appropriateness rather than their general justifiability.

Provisions limiting the scope of advance directives to certain disease states (e.g. coma) or to certain therapies (e.g. ruling out refusal of artificial nutrition) are hard forms of strong paternalism. They constitute a substantial restriction of the right to self-determination and are therefore highly controversial in ethical debate.

1.5 Summary and Conclusion

The overview of historical developments in the US and the situation in various European countries has shown that the importance of advance directives has increased in recent decades. Underlying these developments are, firstly, the new possibilities of (intensive care) medicine and, secondly, a societal change in values, with more weight being attached to the citizen's individual interests and rights than to the interests of the state or of third parties such as physicians or family members.

At the same time, it is evident that opinions still differ widely as to the general authority of advance directives, their scope and the formal conditions for compliance—as reflected in the differing legal regulations of different countries. This is related to some general questions and practical problems concerning advance directives. The answers to these questions again depend on the evaluation of the patient's right to self-determination in relation to other values, such as the duty of the state or physicians to provide care and protect human life. These processes of deliberation are a matter for each individual nation. International regulations of advance directives can thus scarcely move beyond the scope of a minimum demand, such as the Oviedo Convention's policy of taking patients' previously expressed wishes into account if they are no longer able to consent when medical treatment decisions are made.

Acknowledgments I wish to thank Linda Hüllbrock for assistance with translation and Bernd Alt-Epping for constructive criticism.

References

Aigner, G. 2007. Das Patientenverfügungs-Gesetz. Historie und Ausgangslage. In *Das österreichische Patientenverfügungsgesetz. Ethische und rechtliche Aspekte*, ed. U. Körtner, Ch. Kopetzki, and M. Kletecka-Pulker, 74–80. Wien/New York: Springer.
Binet, J.R. 2008. France. In: *Country reports on advance directives*, ed. S. Brauer, N. Biller-Andorno, and R. Andorno, 27–29. Zurich: University of Zurich.
Brauer, S. 2007. Die Patientenverfügung in der Schweiz. *Bioethica Forum* 55: 26–28.
Brauer, S., N. Biller-Andorno, and R. Andorno (eds.). 2008. *Country reports on advance directives*. http://www.ethik.uzh.ch/ibme/veranstaltungsarchiv/2008/ESF-CountryReports. pdf. Accessed 25 July 2012.
Brown, B.A. 2003. The history of advance directives: A literature review. *Journal of Gerontological Nursing* 29(4): 4–14.
Buchanan, A.E., and D.W. Brock. 1989. *Deciding for others: The ethics of surrogate decision making*. Oxford: Cambridge University Press.
Childress, J.F. 1982. *Who should decide? Paternalism in health care*. Oxford/New York: Oxford University Press.
Council of Europe. 1997a. Convention for the protection of human rights and dignity of the human being with regard to the application of biology and medicine: Convention on human

rights and biomedicine. http://conventions.coe.int/Treaty/en/Treaties/Html/164.htm. Accessed 25 July 2012.

Council of Europe. 1997b. Convention for the protection of human rights and dignity of the human being with regard to the application of biology and medicine. Explanatory report. http://conventions.coe.int/Treaty/en/Reports/Html/164.htm. Accessed 25 July 2012.

Cugliari, A.M., T. Miller, and J. Sobal. 1995. Factors promoting completion of advance directives in the hospital. *Archives of Internal Medicine* 155: 1893–1898.

Damato, A.N. 1993. Advance directives for the elderly: A survey. *New Jersey Medicine* 90: 215–220.

Faden, R.R., and T.L. Beauchamp. 1986. *A history and theory of informed consent.* New York/Oxford: Oxford University Press.

Fagerlin, A., P.H. Ditto, J.H. Danks, R.M. Houts, and W.D. Smucker. 2001. Projection in surrogate decisions about life-sustaining medical treatments. *Health Psychology* 20: 166–175.

Feinberg, J. 1971. Legal paternalism. *Canadian Journal of Philosophy* 1: 105–124.

Geth, C. 2010. Passive Sterbehilfe. Basler Studien zur Rechtswissenschaft. Reihe C, Strafrecht. Band 24. Basel: Helbing Lichtenhag.

Inthorn J., and M. Kletecka-Pulker. 2008. Austria. In: *Country reports on advance directives,* ed. S. Brauer, N. Biller-Andorno, and R. Andorno, 5–11. Zurich: University of Zurich.

Kutner, L. 1969. Due process of euthanasia: The living will, a proposal. *Indiana Law Journal* 44: 539–554.

Lehmann, L. 2008. USA. In: *Country reports on advance directives,* ed. S. Brauer, N. Biller-Andorno, and R. Andorno, 103–107. Zurich: University of Zurich.

Meier, D.E., B.R. Fuss, D. O'Rourke, et al. 1996. Marked improvement in recognition and completion of health care proxies: A randomized controlled trial of counseling by hospital patient representatives. *Archives of Internal Medicine* 156: 1227–1232.

Meran, J.G., S.E. Geissendörfer, A.T. May, and A. Simon (eds.). 2002. *Möglichkeiten einer standardisierten Patientenverfügung. Gutachten im Auftrag des Bundesministeriums der Gesundheit.* Münster: LIT.

Parfit, D. 1984. *Reasons and persons.* Oxford: Oxford University Press.

Patzig, G. 1989. Gibt es eine Gesundheitspflicht. *Ethik Medicine* 1(1): 3–12.

Rubin, S.M., W. Strull, M.F. Fialkow, et al. 1994. Increasing the completion of the durable power of attorney for health care: A randomized, controlled trial. *Journal of the American Medical Association* 271: 209–212.

Rurub, M. 2008. The Netherlands. In: *Country reports on advance directives,* ed. S. Brauer, N. Biller-Andorno, and R. Andorno, 53–58. Zurich: University of Zurich.

Schöne-Seifert, B. 1996. Medizinethik. In *Angewandte Ethik. Die Bereichsethiken und ihre theoretische Fundierung,* ed. J. Nida-Rümelin, 552–648. Stuttgart: Kröner.

Seoane, J.A., and Simón, P. 2008. Spain. In: *Country reports on advance directives,* ed S. Brauer, N. Biller-Andorno, and R. Andorno, 83–87. Zurich: University of Zurich.

Simon, A. 2010. Medizinethische Aspekte. In *Patientenverfügungen: Rechtliche und ethische Aspekte,* ed. D. Sturma, D. Lanzerath, and B. Heinrichs, 59–109. Freiburg/Br.: Karl Alber.

Teno, J., J. Lynn, N. Wenger, et al. 1997. Advance directives for seriously Ill hospitalized patients: Effectiveness with the patient self-determination act and the SUPPORT intervention. *Journal of the American Geriatrics Society* 45: 500–507.

Turone, F. 2009. Doctors announce death of woman at centre of Italian 'right to die' case. *BMJ* 338: b574.

Turone, F. 2011. Italian law on advance directives offers no choice for patients. *BMJ* 343: d4610.

Wilkinson, A., N. Wenger, and L.R. Shugarman. 2007. *Literature review on advance directives.* Prepared for office of disability, aging and long-term care policy, office of the assistant secretary for planning and evaluation, U.S. Department of Health and Human Services. http://aspe.hhs.gov/daltcp/reports/2007/advdirlr.pdf. Accessed 25 July 2012.

Zucker, M.B. (ed.). 1999. *The right to die debate: A documentary history.* Westport: Greenwood Press.

Chapter 2
Personal Capacity to Anticipate Future Illness and Treatment Preferences

Marie-Jo Thiel

2.1 Introduction

Advance directives are a new tool whose purpose is to relate end-of-life decisions to autonomy, a core principle of modernity. But since they may be accorded a decisive (binding) role by national legislation, it must also be clear that the individual is able to anticipate future illness and treatment preferences, has understood the associated issues, concerns and challenges, and is willing to proceed with an advance directive, exercising his right to autonomy and fulfilling the legal requirements.

Advance directives allow competent persons to express their wishes concerning end-of-life decisions and to exercise their rights of self-determination in order to influence their future care should they become incapacitated. They generally take the form of a written document (also called a living will, directive to physicians, medical directive or healthcare declaration) that conveys an individual's wishes and preferences regarding medical decisions to be made when one loses decision-making capacity and/or assigns medical power of attorney to a healthcare proxy/ agent. Advance directives have to comply with legal requirements. Although the term *competence* is frequently used interchangeably with *capacity*, it is not exactly the same: "competence" primarily refers to a legal status, involving the ability to understand and participate in legal proceedings.

Since advance directives could potentially play a decisive role when a patient comes to lack decisional capacity (e.g. in coma, severe dementia), certain issues may arise due to the gap between the original statement and the actual decision, to which great weight is attached in the modern world given the importance of autonomy. Is a healthy person able to anticipate? Is it even reasonable to suppose

M.-J. Thiel (✉)
European Centre for the Study and Teaching of Ethics (CEERE),
University of Strasbourg, Strasbourg, France

European Group on Ethics in Science and New Technologies (EGE), Strasbourg, France
e-mail: mthiel@unistra.fr

P. Lack et al. (eds.), *Advance Directives*, International Library of Ethics,
Law, and the New Medicine 54, DOI 10.1007/978-94-007-7377-6_2,
© Springer Science+Business Media Dordrecht 2014

that wishes can be accurately predicted when modern medicine consistently enables real progress in the management of currently fatal diseases? More than ever, personal capacity rests on the historical, situational and philosophical context that increases or decreases the ability to anticipate future illness. It also rests on personal identity and capacity for judgment, which will not be highlighted in this chapter as they are explored elsewhere in this volume, although it is important to emphasize them insofar as they influence personal capacity for anticipation.

We will briefly discuss the role of context and requests in relation to personal capacity. We will examine two examples and finally consider the usefulness of a learning process around advance directives.

2.2   Influence of Context on Capacity for Anticipation

The patient's wishes formally written down in advance directives, as well as the capacity to anticipate future illness and treatment preferences, are based on opportunities and a favourable setting, on values and beliefs. Anticipation is related to what one imagines about the unknown and tries to symbolize from one's fears and expectations. What one imagines is also related to what others imagine, and what they imagine today is not what they imagined yesterday. Over the last 50 years, an unprecedented shift has occurred (Thiel 2010). Analysing the influence of context thus gives us an idea about anticipation as a socially and culturally driven phenomenon.

2.2.1   Historical Context

Historically, advance directives first appeared in the form of a living will as a request for euthanasia—witness the first approach by the Euthanasia Society of America (Olick 2012) and the Quinlan and Cruzan affairs—in a society challenged by the advent of significant but in some ways ambiguous medical progress.

Before that, in a paternalistic environment, the patient was not considered able to understand, still less to anticipate, future outcomes even of a diagnosed disease. Trying to anticipate was wrong, faithless, unsafe and presumptuous. People had to rely on the competence of the physician alone and on the will of God. This emphasizes the cultural and societal construction of capacity to anticipate, which should always be borne in mind.

The modern individual stresses not only the capacity for autonomy but also the necessity to make self-determining and responsible decisions. He cannot see himself as being subjected to death and dying. Can he even speak about it in a society in which death has become a taboo? He seeks by all means to be the subject of his life, including death and dying, in order to control (or deny) it. The modern person wishes to be responsible for his life, actively involved right up to the final stages, and so dying cannot be left to chance. This leads to two attitudes: either advance directives are regarded as necessary to survival, using technology in the way and insofar as one wants to, or advance directives are seen as useless and even "dangerous" in the sense that they could hasten death, and make people feel insecure and hopeless, rather than confident in the promise of medical technologies.

This does not mean that modern individuals easily come to terms with the growing fragility and vulnerability of the body. The task is all the more difficult as they have to face anthropological, as well as technological, shifts, both influencing each other. The modern subject wants to overcome all heteronomous constraints, including his own physical body, which should be spiritualized and digitalized in accordance with the goals of transhumanism and posthumanism, where death is definitively overcome (Thiel 2013).

This perspective does not necessarily contradict the previous one: it extends the central place of reason and reasoning which pushes the modern individual to assume responsibility for his life and death until subjective reasoning is no longer possible. This emphasizes the need to be involved through advance directives. Moreover, it quite often creates the impression of having an immense capacity to anticipate, and the feeling that everything can be mastered—even death. But it overlooks the fact that there is a body behind the reason, which can drive the subject even where he does not want to go, and a society behind the body, which nobody can leave. So we may also have this kind of highly paradoxical anticipation: the virtual body turning against its emotional insignificance. Some may then require the use of technology to the point where it infringes bodily integrity. Others interpret their illnesses, especially cancer, like Fritz Zorn (1976) in his book *Mars*, as a "chance" to get out of this virtual false world and (re)discover physical sensations. How confusing the capacity for anticipation can be!

2.2.2 Situational Context

The capacity for anticipation is as much a matter of self-givenness[1] as givenness-recognition by others. If the capacity I give myself and recognize has to do with my own conception of autonomy (and personal identity), the ability to anticipate also depends on recognition by others, within a "triangular" framework involving the "I" of the patient, his/her family and friends, and the physician; or, more generally, the patient with his/her immediate context, his/her different interlocutors (medical team, proxy, chaplain, spiritual counsellor, friend, professional contacts), and social spheres of belonging (sport, religion, culture…) and the media, which today play a decisive role. All these actors encourage or discourage what a patient thinks about his ability to anticipate and what is foreseeable. Let us consider the role of physicians and collective imagery.

---

[1] The English term *self-givenness* is used to translate the French term *auto-donation*. See, for example, Michael O'Sullivan, Michel Henry. *Incarnation, Barbarism and Belief* (2006), Peter Lang, footnote 5, p. 34.

2.2.2.1 The Role of Physicians

Physicians have contrasting attitudes. Some find the approach of advance directives ambiguous and too vague. Thus, it is hard for them to adhere to such directives, all the more so because they may not feel real empathy, as this is not required by evidence-based medicine. They are reluctant to accept the fact that patients express wishes and choices through advance directives and try to persuade them that advance directives are futile and inappropriate.

Nevertheless, more and more physicians are requesting advance directives since treating a patient without his/her (valid) consent would be unlawful in several countries. When the patient's situation is very difficult, advance directives are welcomed even in those countries where they are not legal tools, as they allow physicians to withdraw or withhold treatments. In such cases, physicians feel "protected", encourage people to consider writing advance directives and are willing to help patients anticipate practical conditions, especially for a diagnosed illness. However, some would agree with the Dutch geriatrician van Asselt (2006) that advance directives are "never available when needed". So the message for a patient sometimes tends to be a condescending double bind: do and do not. This may lead to a vicious circle: the physician does not initiate a discussion on advance directives and the patient does not speak with an unconvinced physician. In the end, no advance directives are drawn up for fear of the patient not being properly treated or understood.

2.2.2.2 Collective Imagery

The individual's surroundings (family, spheres of belonging, media) share collective imagery which may deeply influence the possibility and the content of anticipation. We have already mentioned the role of a society driven by a techno-medical ideology in which there is always something to be done, a technical solution to implement rather than a caring approach which would *also* help people to take into account their own vulnerability and finally their whole life, including their emotions (Thiel 2013).

In fact the collective imagery of medical technologies is highly contrasting: sometimes powerful to the point of prohibiting any anticipation, sometimes utterly fearsome, so that some patients take refuge in their worries (or terrors) rather than really examining the underlying logic; they thus come to express wishes that are not their true desires but rather a bulwark to protect them from techniques and practices viewed as automatically harmful and painful.

We could also say the same about the losses associated with ageing (chronic diseases, disability or excessive dependence): some advance directives are written not as a result of reasoned anticipation, but in order to express a refusal of what is considered senseless and debasing. Later on, people driven by this "counter-reaction" very often revise their position when experiencing real life, in which, as the saying goes, it is better to "measure twice, cut once"; the worst is never certain.

The collective imagery conveys a lot of superficial information, shortcuts and preconceptions. Some low-quality websites or the popular media may contribute to it, conveying "beliefs" that have not been assessed. Relatives may also be equivocal and have a "solution" for every evil, discouraging any anticipation with a "Don't be so pessimistic, have faith in medicine!"; so sometimes they are not able to be faithful representatives of the patient who has become unable to express himself.

Advance directives are often interpreted as euthanasia equivalents; accordingly, some write them only in order to request this possibility and others do not because they are not in favour of this "solution" and/or think it is legally prohibited. Activist groups in favour of euthanasia are behind this imagery and for years have claimed the right to write "living wills", which are almost always requests for euthanasia.

A final and particular emphasis should be put on the administration of fluids and nutrition since they are symbolically, culturally and sometimes religiously expressive of ordinary basic care that should never be stopped, while some laws (including in France) insist that artificial nutrition and hydration are medical treatments which have always to be assessed. It is scientifically documented that when feeding tubes are removed, people in a terminal condition do not feel hungry or thirsty (which are sensations, and not the cause of death). Nevertheless, because eating and drinking play a prominent role in personal and social life, it is necessary to clarify patients' wishes on this point, as collective imagery is deeply ingrained in the mind (Thiel 2008; Comitato Nazionale per la Bioetica 2003).

Finally, if the situational (geographical, cultural, philosophical) context strongly influences the capacity to anticipate future illness and related preferences, it similarly impacts the appointed proxy, who is not necessarily at the same stage as the patient and could be emotionally unable to bear the burdens of future decision-making. This is especially true when episodes of asphyxia occur in the context of terminal respiratory failure or as a complication of a degenerative disease and the patient has clearly refused intubation and clearly informed his proxy and loved ones. It remains difficult to "*do* nothing" (when *doing* is so important) while the patient is dying.

2.2.3 Philosophical Context

Anticipation is directly related to the duty of autonomy as self-determination, which originates from the Enlightenment. But what autonomy? With what consequences for advance directives? And for their author?

Autonomy implies free and informed consent, and an advance directive is primarily anticipatory consent for (non-)treatment. But this principle can be understood in different and even contradictory ways (Thiel and Freys 2008). In most countries, autonomy is an essential aspect of the patient-physician relationship, but in Europe—especially in France, Italy and Germany—there is more room for interpretation than in the US, where the rights of the "sovereign self" have to be absolutely respected. In other words, our lives are ours to shape, and our decisions, once made, should be inviolable. According to Kantian philosophy with its primacy of the good, most Europeans would consider this statement too strong and, above all, related to a requirement of the right which is promoted over the good.

Since the capacity for anticipation is closely related to autonomy, it is understandable that this capacity is understood differently from one country to another. This could provide an explanation for differences in, for example, the number of elderly people with advance directives. Vezzoni (2005), a political sociologist, reports that one in three Americans in nursing homes has written an advance directive, while less than 1 % of the population in the Netherlands have written an advance refusal, and for specific groups such as the elderly, the rate is always below one in ten. In France, the numbers are even lower. A survey of persons aged over 75 carried out by two researchers at the Cochin Hospital Centre for Clinical Ethics in Paris (Eric Favereau et Martha Spranzi 2013) shows that, with the exception of members of the "Association pour le Droit de Mourir dans la Dignité" (ADMD),[2] 90 % are not familiar with the concept of advance directives and, when it is explained, 83 % state they are not interested in it. For 42 %, it is not a bad idea, but too complex, too early, and it does not concern them. For 36 %, advance directives are pointless and 22 % do not want to discuss them.

We should not generalize these data, but take them as pointers and probably as an indication of the need to clarify advance directives and autonomy, and how (willing) people can be made more aware of the challenges and requirements involved. One of these is to ensure that advance directives are free from *undue* influence by others, especially ideological groups, or family members who might be interested in money or insurance, as well as free from other distorting influences (e.g. pain or drugs). Sometimes it could also be prudent—and some legislation (e.g. in Austria) requires this for binding advance directives—to appoint one or two witnesses in order to ensure a free and informed expression of the patient's wishes. These witnesses may also help to clarify wishes and expectations and facilitate the process of anticipation.

Yet what about anticipatory decisions, since these are not actual decisions immediately taken by the autonomous subject? Traditionally, a decision involves three elements (preparation, decision, execution), but there is a degree of contemporaneity (proximity in time and space). In advance directives, however, there is a huge gap between the statement of preferences and the actual decision (if one is indeed taken). Anticipation not only requires a personal capacity but must also try to overcome uncertainties—which are all the greater when a healthy person is stating preferences for (non-)treatment of a hypothetical condition that might (or might not) happen in some possible future. The question then arises whether the uncertainty is so great that, as a matter of ethics, law and public policy, it is reasonable to honour such declarations. This is why some legislation limits the period of validity and/or allows room for interpretation (cf. the Leonetti law in France; see Thiel and Freys 2008).

---

[2] A euthanasia activist group.

The crux of the matter might be the conception of autonomy to be implemented and the room for interpretation of advance directives, which implies different levels of requirements. The issue at stake is indeed the type of society and of values that a community or individual wishes to promote. Autonomy as a right is not promoted in the same way as autonomy as part of an axiological worldview in which vulnerability is assumed to be a fundamental dimension of human beings (Thiel 2013).

Consequently, advance directives may be integrated into decision-making with different requirements for anticipation adapted to the philosophical context, while attempting to clarify the disadvantages of each position. The wish to anticipate future illnesses questions, shifts and reinforces future decisions, in recognizing in the current actor the ability to act without being stopped in his tracks by the prospect of fragility and vulnerability, and by the fear of dying—a fortiori when he knows the pathology he will confront.

Certainly, some will baulk when it comes to writing, some may simply prefer conversation, some remain in denial. Freedom is obtained at this price, but we should probably also keep in mind the role of education. For those who adopt this approach, advance directives "provide a formal framework for the expression of a true word", according to the Genevan theologian Marc Faessler (2005). They require "looking at given situations, to create a climate of trust" (ibid.). They give us the illusion of being free and creative subjects, of having power over disease and death, "a voluntary decision offering salvation, lurking in the depths of consciousness and courageously transforming the world" (Sfez 1984), and of addressing the determinism of impassable finitude, of fragmenting the steps to death, as if to distance oneself from it a little, and to counterbalance it. How indeed is one to accept death that comes irrevocably and inevitably? The feeling of being able to decide something even minimal, to imagine oneself free "anyway", to make oneself helpful in facilitating final decision-making, may sometimes lead to phantasmagorical and illusory perspectives, but can also contribute to rational and progressive acceptance, especially when a serious disease has already taken hold, and to a sense of "living unto death" (Ricœur 2007). The challenge is then both to recognize capacity in everybody and simultaneously to learn autonomy (Autiero 2008) as a way of learning about advance directives and vice versa.

2.3 Personal Capacity

Along with the increasing demand for advance directives, awareness of their difficulties has also grown, and some of their premises have been questioned. The obstacles are related to the context (see above) and also to the ability of individuals to accurately foresee and project themselves into a medical future, to find out and fully assimilate information, and to communicate their wishes and preferences clearly so that advance directives are helpful for future decision-making.

2.3.1   Criticisms Regarding the Medical Situation

Are people able to foresee a medical future accurately? The question is not so much about capacity in itself (everyone has a certain capacity) as it is about the progressive involvement of an autonomous subject (with a history, preferences, etc.) in a decision-making process that starts with *today's* decisions. So anticipating concerns and expressing one's wishes and preferences for the time when one is no longer capable of making decisions is probably a good exercise for the author but is nevertheless fraught with pitfalls.

If a healthy person who has never been confronted with serious illness, either personally or through a loved one, decides to convey advance directives, these are likely to be merely theoretical statements. If the person goes through algorithms, advance directives may be too detailed and confusing, probably with conflicting and/or illogical preferences. In the US, the idea prevails that the more accurate the advance directive, the more binding it could and should be. But there is no evidence for this because the medical circumstances in which one writes an advance directive are quite different from those in which it has to be applied. If the person uses vague and ambiguous terms ("hopeless", in a "terminal condition", with a "lack of quality of life") because it is impossible to foresee all illnesses years in advance, to predict medical progress, and to know about end-of-life care (of which the author has no experience), advance directives will not be sufficiently informed. Some legal statements or ethical opinions mention this issue: "a directive devised decades prior to the need to make an end-of-life decision, and never subsequently revisited, let alone modified, would probably be of little evidential value concerning a person's preferences near the time of death" (Young 2007).

The best solution to minimize the slippage is to reduce the time between the formulation of advance directives and their use, or at least to periodically revise them. Nevertheless, it is essential to understand that one cannot completely eliminate the potential for slippage and it would be counterproductive to try to do so. The threshold would have to be set far too high, ultimately leading to unrealistic demands. It would also indicate a lack of confidence in the (responsible) medical team and in the proxy.

When a serious irreversible disease has begun, people are generally much more receptive and capable of accurately conveying what they want/refuse. Some may still deny or minimize or simply refuse to anticipate anything. But when these patients specify particular practices (e.g. feeding tubes, respirators, dialysis, antibiotics, etc.), advance directives are of high value and must be acknowledged as binding (Virt 2007). Thus, some legislation links the validity of advance directives to elements such as the beginning of a disease, discussions with a physician (and/or lawyer), or a period of 2 or 3 years.

Nevertheless, the start of a terminal condition does not automatically entail the capacity to anticipate its evolution. Patients might be invited (perhaps with others in the same situation?) to enter into (or pursue) a process enabling them to raise their awareness and come progressively to a better understanding of what is going on, including medical progress and significant changes in treatments, in order to integrate hopes or concerns.

However, even then, people often feel unable to tell physicians or write down precisely what they want as regards stopping intensive and life-supporting treatments. Today, numerous studies are seeking to clarify these struggles and to propose tools to overcome the obstacles, especially when advance directives are potentially legal tools, as in the US. However, for some people, as van Asselt (2006) points out, "any consideration of advance directives is inextricable from questions of euthanasia".

Advance directives have to do with the complexity of end-of-life decisions. And the major advances in the medical field and life-sustaining technologies seen in recent years send out *contradictory images*—sometimes worrying, since techniques could bring more suffering, sometimes hopeful, but only if there is an ungrudging "submission" to technology. Most often, these reactions are combined in the same person, so that the ability to anticipate concretely is paralysed by emotions and uncertainty (but not the faculty to go on thinking on the threshold of the conscious/ unconscious).

This makes it essential for physicians (or other healthcare providers) to explain and help people to get rid of the idea that they are "experimental subjects", potentially "hooked up to a breathing machine" and to help them to clarify what they feel progressively able to agree to without despair, so that they can articulate and write down their preferences. Nevertheless, physicians are sometimes reluctant to enter into time-consuming discussions. Hence, preconceptions, confusion and misunderstandings about medical conditions and treatments prevail.

2.3.2 Criticisms Regarding the Individual's Capacity

The individual's capacity is not primarily a matter of intelligence. Anticipation is directly related to what one imagines about the unknown and tries to symbolize from one's own fears and expectations in order to keep as much control as possible. It is inextricably linked to conscious and unconscious layers. It projects consciousness towards the unimaginable while trying to calm fears and the threat of evil. What motivates the advance directive is also what may disturb the individual's perception and communication. And this raises the issue as to whether the role of anticipatory imagination is not totally disproportionate. Capacity indeed is not to be measured, but believed. And beliefs mix rational with metarational and also irrational. But no life is possible outside the belief—probably the most fundamental—that life may be meaningful. Thus, the question of anticipation of future diseases is shifted to the demand for meaning of life when vulnerability and fragility loom large.

The answer brings up the way in which information may interfere with the perception of one's own ability to fight against a disease and stay fit. With the medical diagnosis, the patient still searches for information—especially on the internet, the primary source of knowledge today—but more importantly, he seeks the opportunity to speak about it, to understand it and to clarify what is helpful and what is not worthwhile or even false. Blogs may relay analogous situations but sometimes they also offer very troubling stories, like other types of media (films justifying euthanasia against all odds based on feelings), bogged down in emotions that make some clinical situations tragic and unbearable, very far from reality, hindering the coming discussions with care providers.

Thus, the ability to draw up accurate and cogent advance directives might be thwarted by unconscious preconceptions of some diseases (linked to personal history, experiences and information) or by tendencies to over-/underestimate the impact of infirmities on life. Some overestimate and forget that various infirmities do not affect their entire body, and that several abilities are just waiting to be used creatively and constructively; people very often forget their capacity for adaptation and state in amazement: "I could never have imagined I would be able to do/stand this." And sometimes they even feel better about themselves, especially those with cancer: "I was afraid when I heard about my cancer. I wanted to die immediately, rather than having to consent to being such a wreck. It is a pity to finally have to pass away, but these last 10 years, despite all the hard times I have had to deal with, have been the best in my life. I would never have imagined that." People may be able to adapt to even the most severe disabilities, as shown in the French film *Intouchables* (November 2011) by Nakache and Toledano, which was inspired by the life of Philippe Pozzo di Borgo (quadriplegic since 1993) and his relationship with his home-carer.

Some underestimate and do not (want to) understand they will have to make a decision about, for example, a respirator. Or they know but do not feel able to make a reasoned decision, as if indecision (refuge in uncertainty?) could protect them by giving the illusion that it is not yet the end. This is very often the case in neurodegenerative diseases.

All this emphasizes the necessity to consider advance directives more as a concrete element of a nurturing process than a final document. Advance directives provide an opportunity to think about the future, the meaning of life, but what is true before illness may be false afterwards. People must have the chance to change their mind as their illness progresses because only then do they discover, progressively, the effects and outcomes of the disease and its treatments, along with their own capacities, related to values and available support—and because they only realize the consequences of former directives as their illness progresses. This is also why an advance directive is valid only as long as it has not been invalidated or reformulated, and why European countries often allow some room for interpretation of advance directives so as not to imprison somebody in a position he/she would not assume at this point.

2.3.3 Criticisms Regarding Communication Issues Related to the Patient

2.3.3.1 It Is One Thing Is to Have Wishes in Mind, Another to Communicate Properly

Most advance directives are refusals of—rarely requests for—treatment. This may seem reductive, but in ethics, a refusal is probably far easier to communicate than an approval and generally means a stronger decision (withholding or withdrawal of life-sustaining interventions). Refusals represent the strongest issue, probably not all that a patient wishes to transmit, but what has to be conveyed at all costs. However, if advance directives mean stopping a life-sustaining treatment, are they acceptable? Are they not putting physicians into an impossible position, in the sense that they are required both not to kill the patient and to accept his/her refusal? If each truly autonomous refusal must be respected, advance directives cannot request illegal practices. But sometimes, since advance directives express a person's last wishes, they may put undue pressure on healthcare providers. What can be anticipated also has to be ethically clarified.

2.3.3.2 Can Model Advance Directives Improve Anticipation Capacity?

Besides countries with no regulation, another group (France, Italy) recognizes the value of advance directives (to be "consulted" by the physician) but does not provide an official model (just the obligation to date and sign advance directives). So people are left alone (free) to "imagine" what the end of life could be like. In another group of countries, advance directives have to comply with formal requirements, which in the US differ from one state to another. Sometimes there are standard documents, which, according to Olick (2012), are "typically phrased in legalese", using complex or ambiguous terms that could represent a significant problem for those with limited literacy (and may also fail because they do not accommodate the diversity of religions, cultures, social values, belongings and even languages).

More and more user-friendly forms are now made available by associations, with educational tools to make the advance directive more precise, convenient and accessible. They are widely available, pertinent reminders and may greatly help people to reflect on advance directives.

2.3.3.3 Other Challenges

Even when people are aware and willing, even when advance directive tools/ models are broadly conceived and designed, it remains difficult to summarize one's wishes accurately and anticipate the interpretation process that will occur. The choice of an informed proxy who accepts this role partly resolves this difficulty, insofar as he/she is aware of the author's preferences and continues to discuss them. The combination of conversation and hard copy is a good strategy to communicate personal wishes, and advance directives can be kept in a place known to specific persons, without having to break confidentiality. Nevertheless, the elderly in nursing homes may have nobody except their care providers.

The proxy must be able to make decisions on behalf of the patient and with the medical team—to make the decision the patient would have made in the prevailing circumstances. Thus, this proxy should be very familiar with the patient's views, helping to clarify and anticipate what is wished. Sometimes, this is a difficult double bind. On the one hand, the proxy has to safeguard and vigorously defend positions under the critical scrutiny of the medical team and possibly the family. On the other hand, though familiarity is an advantage, the proxy may be paralysed by emotions. Sometimes he/she is involved in conflicts of interests, since the proxy puts personal preferences to the fore, or death wishes, or conversely refuses to countenance the death of the patient. There is no foregone conclusion.

Finally, given the growing complexity of end-of-life care and decisions, precise anticipations claimed as a strong right are not necessarily more helpful (may well be less helpful?) than more general, value-based wishes as the expression of deep insights and conversations. This emphasizes the role of an advance directive process and education.

2.4 Examples

Advance directives discussed with the physician (when a serious illness has been diagnosed) and with the proxy and loved ones in order to go through essential matters are and remain an ideal that improves anticipation, giving weight to advance directives but without making things obvious. Two examples highlight the ideal and the real situation.

2.4.1 Amyotrophic Lateral Sclerosis

Amyotrophic lateral sclerosis (ALS) is paradigmatic of the difficulties encountered (Thiel and Freys 2008). Normally, patients with this condition remain conscious until death and might not actually need advance directives. However, the progression of ALS may necessitate a tracheotomy. Such patients may live on average for 10–15 years as tetraplegics attached to a respirator, which highlights the need for a mature decision on the wish to be tracheotomized or not. Without written or oral directives, physicians must attempt resuscitation. If the patient refuses resuscitation, healthcare providers have to accompany him/her appropriately. If he/she accepts it, the next of kin must also be ready to accept this decision and to assist the patient for years, since there are very few specialized facilities that can accommodate these patients (Paillisse et al. 2005). And if there is no family? Could a patient living alone really have a desire to live? In this case, anticipation works "perfectly", but is autonomy not ruled out?

Tracheotomy is the greatest concern. All patients know this, as they are treated in specialized units and can first observe the development of ALS in others. But some cannot or do not want to make a decision. They may explain all the issues to the proxy, to their friends and family, they may have perfectly understood, but when it comes to passing from the imaginary to reality, many parameters change, making things difficult for healthcare professionals. A US survey (Albert et al. 2005) showed that only 19 % of patients clearly state a death wish and 6 % express their desire for a tracheotomy. What about the others? The issues are all the fuzzier as there may be a huge discrepancy between the patient, who has gone through his illness, and the family—especially the partner who does not want to lose the loved one. This gap is a source of anxiety and dismay: while a patient accepts the idea of death, next of kin do not understand why not everything is being done to save the patient's life.

What should be the position of healthcare providers? They may be helped by the lucidity of the patient and advance directives, but at the same time anticipation is challenged by uncertainty: nobody knows what this disease will be like in this person, so all actors have to constantly communicate as far as possible. Boldness and prudence should go hand in hand to face the sensitive issue.

2.4.2 Alzheimer's Disease

When healthy people think about Alzheimer's disease or age-related dementia, they are most often afraid and want to write advance directives in order to distance themselves from this black hole, but this can also lead to undesirable decisions. As in ALS, uncertainty is a matter of concern, as is the way the disease is represented, since our society strongly values reason. Alzheimer's disease is usually synonymous with dementia, involving a loss of reason with pain, incontinence, isolation and, for some, a loss of dignity. Thus, in the countries where euthanasia is allowed, this disease could justify such a request. This emphasizes the role played by fears in the anticipation process; it also stresses the importance of advance directives in justifications of euthanasia and underlines the risk of undue pressures that likely lead the elderly to express refusals in advance directives or requests for euthanasia in order to "free" their next of kin (and society).

Anticipation is in fact difficult and very troubling. Even when the disease is diagnosed, it is not possible to predict comorbidities, duration or personal experience. And when an advance directive mentions a refusal of in case of severe dementia, at what point should "severity" trigger withdrawal or withholding of treatment? When pneumonia occurs, must it be treated or is treatment already futile? How may quality of life requirements be interpreted? How can we appreciate the quality of life of demented people? Are we sure we are following the patient's anticipated wishes?

In my listening experience, Alzheimer's disease very often generates great fears, even a feeling of horror, but not always; this is very new and is perhaps linked to better treatment of a disease which is still new, for we have only really been talking about it since the 1990s and we are only now observing reactions against the one-sided language of deficiency: Alzheimer's disease does not necessarily mean meltdown. Thus, some people in their fifties (one of them a psychiatric nurse) facing the age-related disorders of their elderly parents and relatives told me that finally, after deep reflection, they would "like to pass away with dementia since this allows people to die without having to realize they are dying". And, they add, "Inside the Alzheimer's units, people are really recognized, respected and treated according to their dignity." These people, largely in a minority so far, are not unrealistic or unaware of the burden and difficulties, especially when a parent no longer recognizes a child. Yet would they still keep on saying/writing this if they were diagnosed with dementia themselves?

"In the Netherlands", the Dutch philosopher Widdershoven (2012) observes, "euthanasia can be performed in cases where patients do not (or not only) suffer from somatic illnesses, but (also) from mental disturbances, such as chronic mental illness or Alzheimer's disease." And he stresses the difficulty of this issue in his country, where the fundamental criterion for requesting euthanasia is "unbearable suffering":

> This implies specific attention for the due care criteria. How can the physician establish whether the request is well-considered and suffering is unbearable? In an advanced stage of Alzheimer's disease, both the concrete wish and the actual suffering may be hard to interpret. In the case of serious comorbidity, euthanasia may be an option when the patient has an advance directive, and suffering is clearly present. In an early stage, both the wish and the suffering can be discussed between physician and patient. This makes euthanasia an option.

In order to help people to anticipate on the basis of the "right to die" in force in the Netherlands, a Dutch study (Albers et al. 2011) analyses the construct of dignity in order to assess the validity of the patient dignity inventory. The different items usually influencing sense of dignity are assessed as very valuable: they involve physical, psychological, social and existential aspects: not being able to carry out daily tasks or to think clearly, feeling like a burden to others, etc.

Nevertheless, is such a grid about impairments a good tool for appreciating what people think about themselves? Are they not reduced to their deficiency and driven to despair and a feeling of indignity, "unbearable suffering", since these elements are only projections designed to decide about hypothetical end-of-life issues? Is "unbearable suffering" not (sometimes) a pretext that dare not speak its name?

2.5   Utility of a Learning Process

Personal capacity to anticipate future illness is all the more effective when people are helped to understand issues and challenges and are given time to focus not only on medical issues (or the precise concerns of a diagnosed illness) but also on values, (cultural/religious) references, leisure, prized relationships, etc. which could help them to make a decision based on who they are and what they wish (Debout 2005). Various tools now available can help to raise awareness and improve the anticipation capacity through a learning process (which could sometimes be helped by behaviour change programmes).

Since individuals are in a constant state of becoming themselves, since life passes through stages, the value and the validity of advance directives are not only a question of (assessed) anticipation capacity but also of having the opportunity to nurture them when faced with the vagaries of life, in discussions with others and when reading. Three kinds of tools may be helpful and would be even more so if death were not a taboo, and if it were easier to resist the injunction of technomedicine that prohibits consideration of death.

2.5.1 Assessment of Capacity to Consent to Treatment

The Assessment of Capacity to Consent to Treatment (ACCT) approach was developed by Moye et al. (2007) for patients with neurocognitive or neuropsychiatric illness and applied to Alzheimer's disease and schizophrenia. Its insights might also be valuable for evaluating and improving personal capacity for anticipating future illness and treatment preferences.

The tool starts with an interview to clarify values and preferences relevant to medical decisions and highlights four elements according to four standards for decisional incapacity: understanding, appreciation, reasoning and communicating a choice. Hypothetical vignettes are used as examples which help the patient to project in different kinds of situations and finally to specify the four points; these could, for our purposes, represent a kind of grid supporting conversations, since they are often threatened by denial, or confusingly mixed.

We use these elements freely: they are highly effective in the promotion of advance directives at the beginning of a serious disease, as well as in the promotion of an intergenerational ongoing conversation about caring, especially with professional carers and physicians.

1. *Understanding* means "the ability to comprehend diagnostic and treatment-related information", to tell what the risks and benefits are, possibly with the help of a paper list. Especially when patients are already affected by serious illness, it is important to make sure they have understood at least the main elements of the usual progression of this illness, what treatments are effective, what the patient is supposed to do and what happens if treatment is refused.
2. *Appreciation* focuses on *understood* information and means "the ability to relate the treatment information to one's own situation, in particular, the nature of the diagnosis"—the outcomes, doubts and hopes associated with adopted treatments, the *personal* consequences and the concerns for family and loved ones.
3. *Reasoning* is "the ability to provide rational explanations or to compare treatment alternatives in a logically consistent manner", to integrate uncertainty and rational reasons (pros and cons) justifying choices which are as consistent as possible. Reasoning comes back to appreciation in order to explain how far preferences depend on personal beliefs and values.

4. *Communicating a choice* means "the ability to convey a treatment choice" or refusal, to make an end-of-life decision, to tell it to another person (proxy, healthcare professional), to write it in the form of advance directives, communicating what has slowly matured in se and in relation to others. This might be seen as an ultimate stage, but it should not because as long as a patient is alive, wishes are likely to be readjusted.

This approach not only "measures" ability but also incorporates the role of context, values, preferences and (cultural, religious) references which might help people to go further in reasoning about the end of life. The interview should focus on three sets of values:

1. *Valued activities and relationships.* Is this individual used to doing sport, music, painting and enjoying hobbies? Is he/she restricted in these activities? Is taking care of him-/herself or not being dependent on others for help in daily life something valuable? What relationships are ongoing with family, friends, communities, associations?
2. *The individual's preferred style for decision-making* (autonomous, shared, deferred): when an important decision has to be made, especially healthcare decisions, what strategy is usually adopted: a decision only made by myself? With a little help from others (who)? Together with…? On behalf of…? The decision is entirely made by… This point is all the more convincing as professionals are constantly tempted to constrain people to act as autonomous subjects writing advance directives since they are not used to functioning in this way.
3. *The individual's beliefs and views* about what should happen if… Is living as long as possible more important than quality of life? Or conversely, is quality of life most important, even if this would shorten life somewhat? How much do religious or spiritual beliefs influence the decision to be made? For example, is the decision entirely deferred to God since His "will has to be done", or to a priest in the name of God, or do religious beliefs influence the decision only slightly or not at all?

The fundamental point is that a conversation is initiated and goes on as a learning process, which is "recorded" as advance directives or otherwise at some points. A lot of tools thus use "values history" in order to create a memory of a person that can be recalled at times of medical need.

2.5.2 A Computer-Based Decision Aid

Since computer use has become more and more widespread, some authors are using this tool to help people, especially in the process of advance care planning. But it can also be used to raise awareness or assess capacity in advance directives. Levi and Green (2010) have imagined an interactive, self-directed computer program, *"Making Your Wishes Known: Planning Your Medical Future"*. It uses a question-answer format involving audio, text, graphics, patient vignettes and videotapes by "professional experts". According to the authors, "this program takes an educational approach, simulating the kind of idealized discussion one might have with an experienced, reflective health care professional who is well informed and has ample time"—precisely what is so difficult in real life. This program, a pilot version of which is presented in the article (available at https://www.makingyourwishesknown.com/), is a truly flexible and highly educational tool based on multi-attribute utility theory (MAUT) and adjusted to accommodate special needs, to help translate values and goals into meaningful advance directives while serving as a reminder for designated proxies. Saved in an online database, a copy goes to the author, who can also send it to the mandated persons. This approach makes people enter a learning and capacitation process. Users are "encouraged to revisit the decisions outlined in their advance directive" and to regard it "as a starting point for initiating substantive discussions with loved ones and healthcare providers".

Obviously, only motivated individuals used to computers can fully apply this program, but there is no need to have prior experience with computers. MAUT tracks inconsistency in the answers and asks for clarification if needed. Moreover, this program can potentially enrich users lives, by also stressing a number of positive statements (about their goals and wishes, the way they see their end-of-life experience...) they can share with friends or next-of-kin.

2.5.3 Enabling Through Behaviour Change

While very few people express their wishes spontaneously through advance directives, the "transtheoretical model of health behavior change" developed by Prochaska (1997) has produced some spectacular results—e.g. at the Gundersen Lutheran Medical Foundation, where 85 % of patients have advance directives. Prochaska's approach, supported by numerous studies, shows that people go through a variety of stages before finding out the behaviour they want to adopt, especially over the long term, with regard to end-of-life decisions. This dynamic process involves six stages, with interpersonal communication based on three pillars: (1) vocational training, (2) patients' training, (3) alignment of organization. The ultimate aim is to avoid making end-of-life issues taboo, to help people express themselves and to emphasize that self-determination through advance directives cannot be conveyed instantly on a whim.

In the *precontemplation stage*, people have no plans or wishes for the foreseeable future and are generally unaware of possible specific benefits: they have no /not enough/false information or they have previously failed; sometimes they fear being confronted on this sensitive topic. The objective is thus to encourage them to ask friends or family why they act and to gradually discover the benefits of proceeding.

In the *contemplation stage*, people are aware of the pros and cons of changing, and this may produce profound ambivalence that can keep people stuck in this stage for long periods. The goal is to get these people to identify the roadblocks, to overcome the hurdles and to make a commitment. If no positive representation can be mobilized, no real and profound engagement will be possible.

When people reach the *preparation stage*, they are ready to express their end-of-life wishes and to write them down imminently, but they often need to be supported with an action plan and a mediator in order to share what is going on with family and friends.

The next stage is logically *action*: advance directives are written, a proxy is requested and accepts the mission. This stage is an achievement even if it requires a lot of energy and commitment. But it is not the last stage, which is *maintenance*. People must be reassured about their decision, about ways to make their wishes known to the next of kin, care professionals and so on. Changes can be made at any time and revisions may be considered at defined intervals, even though experience shows that people almost never come back to modify substantially such mature advance directives. Such a process also helps people not to focus on perceived dignity at a difficult time, when a serious disease has been diagnosed and is considered to undermine self-esteem, which could potentially lead to a request for euthanasia in countries which have legalized it.

2.6   Conclusion

Advance directives are a tool that should be neither treated in an absolutist manner nor devalued. Each person is called on to make autonomous and responsible decisions and has a certain capacity to anticipate future illness and treatment preferences, but this also has to be improved. Thus, an ongoing learning process remains the core challenge for building anticipation capacity.

References

Albers, Gwenda H., H. Roeline, W. Pasman, et al. 2011. Analysis of the construct of dignity and content validity of the patient dignity inventory. *Health and Quality of Life Outcomes* 9: 45–54.
Albert, S.M., J.G. Rabkin, M.L. Del Bene, et al. 2005. Wish to die in end-stage ALS. *Neurology* 65: 68–74.
Autiero, Antonio. 2008. Patientenverfügung und Patientenautonomie: Plädoyer für eine ethische Pädagogik. *Bulletin de la Société des Sciences Médicales du Grand-Duché de Luxembourg* 3: 311–327.
Comitato Nazionale per la Bioetica. 2003. *Dichiarazioni anticipate di trattamento*. Roma. http://www.governo.it/bioetica/testi/Dichiarazioni_anticipate_trattamento.pdf. Accessed Apr 2012.
Debout, Christophe. 2005. La contribution infirmière dans le recours aux directives anticipées. *Soins* 708: 38–40.

Eric Favereau et Martha Spranzi (ed.). 2013. *Les directives anticipées chez les personnes de plus de 75 ans*. Paris: Centre d'éthique clinique, Hôpital Cochin. Ed. de l'Assistance publique des Hôpitaux de Paris.

Faessler, Marc. 2005. L'enjeu spirituel des directives anticipées. *Rev Internat de soins palliatifs* 20(4): 135–137.

Levi, Benjamin H., and Michael J. Green. 2010. Too soon to give up: Re-examining the value of advance directives. *The American Journal of Bioethics* 10(4): 3–22.

Moye, Jennifer, Michele J. Karel, Barry Edelstein, et al. 2007. Assessment of capacity to consent to treatment: Challenges, the 'ACCT' approach, future directions. *Clinical Gerontologist* 31(3): 37–66.

Olick, Robert S. 2012. Defining features of advance directives in law and clinical practice. *Chest* 141: 232–238.

Paillisse, C., L. Lacomblez, et al. 2005. Prognostic factors for survival in amyotrophic lateral sclerosis patients treated with riluzole. *Amyotrophic Lateral Sclerosis and Other Motor Neuron Disorders* 6: 37–44.

Prochaska, James O. 1997. The transtheoretical model of health behavior change. *American Journal of Health Promotion* 12(1): 38–48.

Ricœur, Paul. 2007. *Vivant jusqu'à la mort Suivi de Fragments*. Paris: Seuil.

Sfez, Lucien. 1984. *La décision*. Paris: PUS (4e éd., 2004).

Thiel, Marie-Jo. 2008. Hydratation et alimentations artificielles en fin de vie. *Revue des sciences sociales* 39: 132–145.

Thiel, Marie-Jo. 2010. Human dignity: Intrinsic or relative value? *Journal International de Bioé thique* 21(3): 51–62.

Thiel, Marie-Jo. 2013. La corporéité face à la maladie et la mort. In *Exploring the boundaries of bodiliness. Theological and interdisciplinary approaches to the human condition*, ed. Sigrid Müller et al. Göttingen: Vienna University Press.

Thiel, Marie-Jo, and Guy Freys. 2008. Les directives anticipées en France. Réflexion éthique. *Bulletin de la Société des Sciences Médicales du Grand-Duché de Luxembourg* 3: 311–327.

van Asselt, Dieneke. 2006. Advance directives: Prerequisites and usefulness. *Z Gerontol Geriat* 39: 371–375.

Vezzoni, C. 2005. The legal status and social practice of treatment directives in the Netherlands. Thesis, University of Groningen, 199p. http://dissertations.ub.rug.nl/faculties/jur/2005/c.vezzoni/. Accessed Apr 2012.

Virt, Günter. 2007. Ethische Begründungen und Kriterien für Patientenverfügungen. In *Damit Menschsein Zukunft hat*. Würzburg: Echter Verlag.

Widdershoven, Guy A.M., and Ron L.P. Berghmans. 2012. Euthanasia and palliative care for dementia patients in the Netherlands. In *Ethical challenges of ageing*, ed. M.J. Thiel. London: Royal Society of Medicine.

Young, Robert. 2007. *Medically assisted death*. New York: Cambridge University Press.

Zorn, Fritz. 1976 (French version 1982). *Mars*. Paris: Gallimard Folio.

Chapter 3
Advance Directives in Psychiatry

Jochen Vollmann

3.1 Introduction

Legal regulations for advance directives exist in many Western countries. In Germany, for instance, a law on advance directives came into force in 2009. In day-to-day practice, such regulations provide binding guidance for all parties and greater legal certainty (Lipp 2009). Democratic legitimation ensures that regulations on advance directives do not merely serve the interests of particular groups or individuals. In connection with such regulations, important and often controversial problem areas need to be considered and discussed from an interdisciplinary perspective in the legislative process. These include the following ethical issues (Sass and Kielstein 2001; Nationaler Ethikrat 2005) addressed by the legislation in Germany (with similar regulations existing in other Western countries):

- Authority
- Scope
- Requirements for validity
- Implementation

The patient's self-determined wishes expressed in an advance directive are binding and must be respected by the physician and other persons. The prerequisite is that the patient has written the directive voluntarily and with the capacity for self-determination, or competence to give consent. Furthermore, the treatment

---

Parts of this chapter were originally published in: Vollmann J: Patientenverfügungen von Menschen mit psychischen Störungen. Gültigkeit, Reichweite, Wirksamkeitsvoraussetzung und klinische Umsetzung [Advance directives in patients with mental disorders. Scope, prerequisites for validity and clinical implementation] *Der Nervenarzt* 83:25–30 (2012).

J. Vollmann (✉)
Institute for Medical Ethics and History of Medicine, Ruhr-Universität Bochum,
Markstraße 258a, 44799 Bochum, Germany
e-mail: jochen.vollmann@ruhr-uni-bochum.de

preferences for particular medical situations recorded in the advance directive must be as specific as possible and be applicable to the current situation. Advance directives are then more than merely an aid for physician decision-making or for determining the patient's presumed wishes (*authority*). Advance directives are not restricted in *scope*, i.e. they are valid regardless of the type and stage of disease. There is, explicitly, no restriction of the scope of advance directives—as debated in the past—to the end of life, to an unfavourable prognosis or to terminal disease, etc. The *requirements for validity* specified in the legislation are that an advance directive is to be prepared in writing and signed by the patient, who is to be of legal age. In Germany (as in many other countries), a prior mandatory consultation with a physician or notarization is not required. The *implementation* of advance directives places the focus on the process of dialogue between the physician and the healthcare proxy or legal guardian. In clinical practice, decision-making is to be facilitated by professional communication among these parties. As in many other countries, the guardianship court is only to be involved in the event of disagreement among the above-mentioned parties in situations where there is a serious risk to the health or life of the patient.

The aim of the Law on Advance Directives is to enable autonomous individuals to record, in a binding form, self-determined decisions concerning future medical treatment in situations where they lack the capacity for self-determination. Many individuals thereby wish to avoid unnecessary suffering, which they may have experienced, for example, in previous episodes of illness or past hospital treatments, etc. Other motivations are to ease the burden for family members and attending physicians in making difficult treatment decisions in the event of the patient's incapacity. Advance directives are binding if they contain information on treatment preferences applicable to the current health situation. The nature and extent of the treatments desired or refused should be indicated as specifically as possible. Because this is not always possible in advance, it is helpful to provide general information on one's personal values, religious beliefs and attitudes towards life and quality of life. Some physicians object that previously expressed self-determined wishes can no longer be applicable for a patient who lacks the capacity for self-determination, since clinical experience shows that a patient's wishes often change during the course of a disease. However, this objection is not valid: the purpose of the advance directive is precisely to enable individuals—at a time when they are capable of making treatment decisions—to give binding instructions for these medical situations in advance. The patient's wishes specified in the advance directive then take precedence over the presumed "natural will" in the concrete medical situation. Moreover, it is difficult to imagine how the self-determined values expressed in the patient's treatment choices could subsequently change at a time when the capacity for self-determination has been lost (Emanuel et al. 1994; Vollmann 2000a; Wittink et al. 2008). Accordingly, as long as the above-mentioned requirements for validity are met, a psychiatrist treating a patient with dementia who is incapable of giving consent must respect the self-determined wishes specified in the advance directive rather than current expressions of "natural will" (gestures, posture, mood, general health condition, etc.). In clinical practice, this can lead to difficult medical-ethical decision situations. Every individual should therefore understand the legal force of an advance directive and the associated consequences. An increase in patient self-determination always entails increased assumption of responsibility by patients. There is thus a need for individual information and counselling, which in practice has not been widely available to date owing to a lack of counselling programmes and expertise.

3.2 Advance Directives for Patients with Mental Disorders

Even though advance directives have primarily been discussed in relation to end-of-life treatment for patients with severe somatic diseases such as cancer (Sass and Kielstein 2001; Burchardi et al. 2005a, b; Lang-Welzenbach et al. 2005; Nationaler Ethikrat 2005; for overviews see Vollmann and Knöchel-Schiffer 1999 and Vollmann and Pfaff 2003), the scope of directives is not restricted either in Germany or in a number of other countries. This means that advance directives are valid regardless of the type and stage of disease, including all forms of mental illness (cf. Deutsche Gesellschaft für Psychiatrie 2009, 2010; Finzen 2009; for a legal perspective see Olzen et al. 2009).

In Germany, as elsewhere, advance directives have thus far played only a limited role in psychiatric practice (Haupt et al. 1999; Vollmann 2000a; Fritze and Sass 2003; Lauter and Helmchen 2006; Hansen et al. 2008). However, an online search using the terms "advance directive" and "psychiatry" turned up numerous websites and forums of interest groups for the mentally ill and people with experience of psychiatric treatment, as well as counselling programmes offered by associations, hospitals, lawyers, etc. Here, the points most frequently mentioned in relation to advance directives for people with mental disorders were as follows:

- general refusal of treatment, e.g. with antipsychotics, electroconvulsive therapy (ECT), occupational therapy
- specific refusal of treatment with high-potency antipsychotics
- specific refusal of treatment with certain (other) psychotropic agents
- specification of a maximum dose
- specification of a maximum duration of treatment
- specification of an antipsychotic-free treatment interval
- approval of treatment under specific conditions (e.g. transfer to an open ward)
- approval of treatment in specific situations (e.g. acute suicidality)

In the US, patients and healthcare providers have had many years of experience with advance directives in psychiatry (Appelbaum 1991, 2004; Swanson et al. 2006a, b; Swartz et al. 2006; Honberg 2010; National Resource Center on

Psychiatric Advance Directives 2010). Empirical studies show that the content of advance directives most frequently concerns the following areas (Srebnik et al. 2005):

- preferred medications, most often antidepressants and second-generation antipsychotics (listed by 81 % of study participants)
- undesired medications, especially first-generation antipsychotics (64 %)
- preferred alternatives to hospitalization (68 %)
- methods of de-escalating psychiatric crises (89 %)
- refusal of ECT (72 %)

In addition to the advance directive, 46 % of those surveyed had appointed a surrogate decision maker in case of mental illness. Fifty seven percent of the participants explicitly stated that they wished their directive to be irrevocable during periods of incapacity (Srebnik et al. 2005).

In contrast to the frequently expressed reservations and scepticism of psychiatrists towards advance directives, empirical studies from the US paint a more positive picture: for 95 % of advance directives, the patients' treatment preferences were rated by psychiatrists as feasible, useful and consistent with practice standards (Srebnik et al. 2005). Thus, advance directives can play a constructive role in psychiatric practice and influence physicians' treatment decisions in line with patients' wishes (Wilder et al. 2007). Another study showed that successes in treatment can be achieved by educating and informing patients about psychiatric advance directives. Facilitated sessions increased the number of advance directives completed by patients. The specific information on treatment preferences given in the directives was rated by psychiatrists as consistent with standards of community practice (Swanson et al. 2006b).

From an ethical perspective, it can be noted that psychiatric advance directives are used in practice and are desired and respected by patients and psychiatrists alike. There is no fundamental ethical difference between the applicability of advance directives for mental as opposed to somatic disorders.

3.3 Capacity for Self-Determination

A requirement for the validity of advance directives is the patient's capacity for self-determination at the time of writing. Since mental disorders can compromise the patient's self-determination more frequently than somatic illnesses (Grisso and Appelbaum 1998; Vollmann et al. 2003, 2004), the assessment of capacity at the time when an advance directive is written merits special attention in psychiatric practice. Terms used in practice include "of sound mind" and "in full possession of one's mental faculties", as well as ability to manage one's financial affairs, competence to consent or capacity for self-determination, capacity for judgement and decision-making, etc.; however, the definition of these terms often remains unclear.

Capacity for self-determination requires the fulfilment of the following criteria:

- Ability to understand information
- Ability to reason
- Ability to appreciate a mental disorder
- Ability to appreciate the possible benefits of treatment
- Ability to make a decision and express a choice

The *ability to understand information* means that information given by a psychiatrist in an effort to obtain informed consent is understood by the patient. To prevent the patient from merely repeating what has been said without understanding it (which would be merely a test of memory rather than understanding), it is advisable to ask the patient to recapitulate the information in his or her own words. It is important that the information given is actively comprehended by the person concerned. The required ability to understand information applies not only to the communication of medical information by the psychiatrist, but also for all other areas.

Having understood the information given, the patient must integrate it into the context of his or her life and system of values in order to be able to make a self-determined decision. The patient must assess what this medical information means for his or her life planning, preferences and values. The advantages and disadvantages of the proposed treatment options need to be weighed up and assessed with respect to the consequences and alternatives. It is precisely this individual assessment process which enables self-determined decision-making (*reasoning*).

Patients with the capacity for self-determination are also able to recognize that they have a mental disorder (*appreciation of the disorder*) and that there are treatment options which could help (*appreciation of treatment benefits*). It is explicitly not required that the patient must personally accept a medical diagnosis (e.g. schizophrenia) or psychiatric conception, or use this terminology him/herself. Rather, for the assessment of capacity, what is important is that patients recognize that they are affected by a mental disorder and acknowledge potential treatment options. Finally, patients must be able to *make a decision and express a choice* (Grisso and Appelbaum 1998; Vollmann 2000b).

In the international psychiatric literature, the MacArthur Competence Assessment Tool (MacCAT) is considered to be the gold standard for assessing decisional capacity (Grisso and Appelbaum 1998; German translation in Vollmann 2008). Tools are available for the areas of treatment (MacCAT-T), clinical research (MacCAT-CR) and criminal adjudication (MacCAT-CA). MacCAT-T formed the basis for the development of the Decisional Competence Assessment Tool for Psychiatric Advance Directives (DCAT-PAD). This tool assesses patients' basic grasp of advance directives, understanding of content, and reasoning abilities (Elbogen et al. 2007). Specifically, it evaluates their ability to:

- understand the key components of psychiatric advance directives (PADs)
- appreciate whether or not PADs would be relevant to them and their treatment
- reason about how PADs would affect their lives

- choose whether they would want to fill out a PAD
- understand the pros and cons of hospital treatment
- appreciate whether hospitalization may be a relevant option for them
- reason about how hospital treatment would affect their lives
- choose whether they would want to be hospitalized if they became ill

Professional assessment and documentation of the capacity for self-determination when an advance directive is written of great significance for the ethical and legal validity of the patient's wishes recorded therein. For unlike mere expressions of the patient's natural will, a self-determined expression of wishes imposes an ethical and legal obligation on third parties (e.g. physician, family members) to respect the patient's wishes, even when these run counter to the physician's advice or a medical indication for treatment.

3.4   Problems with Assessments of Capacity

In medical ethics, the principle of self-determination obliges physicians to obtain a patient's consent prior to medical treatment (Beauchamp and Childress 1994). A condition for valid informed consent is the patient's competence, which may be impaired, particularly in patients with mental problems (Helmchen and Lauter 1995; Koch et al. 1996; Helmchen 1998; Vollmann 2000b).

In clinical practice, physicians generally presume competence on the patient's part. But if they have occasion to assess competence, they will generally proceed using their own subjective judgement and clinical experience and have difficulty applying standards suggested in the literature (McKinnon et al. 1989; Markson et al. 1994). Physicians also often make inconsistent evaluations of competence in a given case (Marson et al. 1997). Considering the ethical and legal significance of competence and the desire for physicians' evaluations to be transparent and reliable, various objective testing procedures have been developed and applied in clinical trials in recent years (Janofsky et al. 1992; Bean et al. 1994; Marson et al. 1995; Kitamura et al. 1998; for an overview see Bauer and Vollmann 2002).

In studies using structured test interviews, a considerable proportion (40–75 %) of acutely psychotic and schizophrenic patients had no competence to consent (Grossman and Summers 1980; Appelbaum et al. 1981; Hoffman and Srinivasan 1992; Grisso and Appelbaum 1995). Patients with depression, however, displayed fewer impairments in empirical studies (Grisso and Appelbaum 1995; Appelbaum et al. 1999). Among patients with dementia, the proportion of non-competent persons increased both with increasing severity of disease and with increasing stringency of the legal standards applied (Marson et al. 1995; Kim et al. 2001). A significant correlation between competence and multiple cognitive functions has been described in demented patients (Marson et al. 1996).

In only a few studies has competence been directly compared among various diagnostic groups using the same test instrument. One such study, the MacArthur

Treatment Competence Study (Grisso and Appelbaum 1995), compared schizophrenic, depressed and medical patients using instruments related to the following four legal standards:

1. ability to understand information relevant to the decision about treatment;
2. ability to manipulate the information rationally (or reason about it) in a manner that allows one to make comparisons and weigh options;
3. ability to appreciate the significance for one's own situation of the information disclosed about the illness and possible treatments; and
4. ability to express a choice.

Patients with scores below defined limits were categorized as impaired in that standard. The majority of patients in all diagnostic groups performed adequately in all standards. However, patients with schizophrenia—as a group—had significantly more deficits than the other two diagnostic groups, and this group displayed the largest proportion of patients with impairment. A hierarchical order of the standards was not detected. For each patient group, a combination of standards resulted in an increase in the proportion of impaired patients, with impairment being defined as poor performance on any of the standards. Since the instruments used were too cumbersome for application in day-to-day clinical practice, the authors developed a short form with a high degree of inter-rater reliability, the MacArthur Competence Assessment Tool-Treatment (MacCAT-T) (Grisso et al. 1997; Grisso and Appelbaum 1998).

Using different standards of competence, substantial differences were found among patients in various diagnostic groups (Vollmann et al. 2003). Studies showed that schizophrenic patients had greater impairments than depressed patients in the standards relevant for competence to consent to treatment (Bean et al. 1994; Grisso and Appelbaum 1995). Patients with dementia showed even greater impairments than schizophrenic or depressed patients. For each standard, a significantly greater proportion of patients with dementia was impaired than in the other two diagnostic groups (Vollmann et al. 2003). Despite the substantial proportion of impaired patients, many patients displayed no impairments in the standards used; thus, the diagnosis is not suitable as the sole indicator of incompetence in individual cases.

The paradoxical situation observed among many schizophrenic patients, who generally appreciated the treatment benefit despite reduced appreciation of the disorder, reminds us of the "double accounting" among patients with schizophrenic psychoses, familiar from clinical practice. The conspicuously high proportion of such patients with full appreciation of treatment benefit may possibly be explained by a selection effect, since patients with a negative attitude towards treatment frequently also refuse to participate in clinical studies.

For a thorough assessment of competence in practice, all standards should be examined, since deficits in just one standard can call overall competence into question. However, the more thoroughly competence is examined (i.e. the more standards are used in assessing it), the more patients will be evaluated as incompetent (Vollmann et al. 2003). The selection and combination of standards depend on previous value judgements. This influences both the content requirements and the threshold for defining incompetence to consent.

Compared to the patients categorized as unable to consent by clinical assessment, the proportion of patients impaired in at least one standard of the MacCAT-T is significantly higher. The discrepancy between objective testing methods and clinical assessment in evaluating competence has been described in the literature (Rutman and Silberfeld 1997; Vollmann et al. 2003). The question arises of whether clinical assessment or objective testing methods are more suitable for satisfying the ethical demand to respect patients' self-determination and also to protect the well-being of patients with impaired competence (and safeguard them against serious and dangerous consequences of incompetent decisions). In the light of the discrepancy found between clinical assessment and objective testing, we must decide whether the risks of decision substitution for possibly competent patients (objective testing) are greater than the risks of possibly incompetent patients making their own decisions (clinical assessment) (Vollmann 2000c).

The MacCAT-T showed good applicability in clinical practice. Most patients evaluate the interview as positive since it gave them the opportunity to discuss their illness and possible forms of therapy at length with a physician. In our experience, patients have great communication needs in this regard. However, some patients complained after the interview that it required a lot of concentration and was difficult at times. For instance, one patient with her first manifestation of schizophrenia said that the test was too much for her on the day of admission (Vollmann et al. 2003). This patient's statement illustrates the practical significance of informed consent as an "educational process" (Roth 1983) and evaluation of competence as an "evolving process influenced by therapeutic interventions" (Mahler and Perry 1988). Reduced abilities related to competence in schizophrenic patients can be improved by educational interventions (Carpenter et al. 2000). More intensive education strategies tailored to individual patients are of great importance in this regard.

There are difficulties administering the MacCAT-T to patients with severe dementia who have major cognitive impairments. This raises the question of whether, as critics suggest (Elliot 1997; Charland 1998; Welie 2001), the criteria for objective testing methods are too strictly or too one-sidedly oriented towards cognitive functions, while ignoring evaluation of major emotional factors. The identification of criteria used for decision-making by people with cognitive impairments (e.g. emotional, social-context-specific and biographical), and the integration of these criteria into the evaluation of competence, may be starting points for further research (Breden and Vollmann 2004). In view of demographic developments—with an increasing proportion of older people in the population, and hence also of patients suffering from dementia—such research approaches are of considerable practical significance.

Controversy surrounds the question of whether objective testing methods can replace clinical assessment of competence. A major problem lies in setting the cut-offs in the tests. In many studies, these are set on the basis of statistical considerations. Studies comparing objective testing methods with other evaluation methods (clinical assessment, forensic-psychiatric study) range from complete agreement (Janofsky et al. 1992) to significant discrepancies (Rutman and Silberfeld 1997). In our opinion, cut-offs have the primary purpose of allowing statistical comparison of groups and illuminating deficits in examined standards; a categorical decision as to whether competence is present or absent cannot be made solely on the basis of the test results. We view the MacCAT-T as a suitable instrument for detecting deficits in patients' decision-making abilities in a concrete case; it should be followed by a thorough clinical evaluation that also includes non-cognitive aspects.

3.5 Ethical Conclusions

Advance directives are valid for the treatment of all illnesses, regardless of whether they are somatic or mental. The considerations presented here refer exclusively to the treatment of patients with mental disorders. The special cases of court-ordered commitment to inpatient treatment and coercive measures for individuals posing a danger to others cannot be discussed further here (cf. Olzen et al. 2009).

The prerequisite for the validity of an advance directive is the patient's capacity at the time when it is written. Since patients with mental disorders may have lost the capacity for self-determination, special attention must be paid to the assessment and documentation of this matter. This requires a clear ethical and legal definition of capacity for self-determination or competence to give consent. Apart from the question of capacity, important points are how closely the situation envisaged in the advance directive corresponds to the current medical situation and the specificity of the treatment preferences expressed. Since, in the nature of the case, the criterion of a close fit between the advance directive and future situations cannot always be met, it is helpful to include more fundamental information on values, the patient's biography, etc. It is further recommended to combine the advance directive with the designation of an informed and trusted individual via a durable power of attorney for healthcare.

Neither the severity of disease nor a conflict between the self-determined wishes and a medical indication for treatment is sufficient to override the wishes expressed in a valid advance directive. In the event of a conflict, the patient's self-determined, closely fitting wishes prevail over specialist medical advice. The physician's duty to provide medical care, a medical indication or a psychiatric patient's fundamental right to receive medical treatment cannot be taken to constitute a duty to treat, outweighing the patient's self-determined wishes. This also applies to life-threatening diseases, as shown by the extensive discussion in medical ethics and medical law with regard to somatic illnesses.

These clinical conflicts and ethical dilemmas have been clearly resolved by legislators in favour of the patient's self-determined wishes. This is demonstrated by examples and experience in somatic medicine, such as the fundamental refusal of blood transfusions on religious grounds by patients who are Jehovah's Witnesses, even in life-threatening situations. By analogy, this also applies to life-threatening situations in psychiatry, such as the self-determined refusal of ECT as a medically indicated and life-saving measure in pernicious catatonia. The prerequisite for the legal force of the patient's wishes, however, is that they are self-determined in accordance with clearly defined criteria, as opposed to mere expressions of natural will. This means that the patient has been adequately informed about his or her health situation, treatment options (nature, purpose, advantages and disadvantages, alternatives) and prognosis. If the patient's guardian and the attending physician disagree in their interpretation of the advance directive, the guardianship court must be called on to make a decision if there is a serious risk to the patient's life or health.

The more authentically the patient's self-determined wishes are expressed, the more helpful and convincing an advance directive will be. This applies particularly to the expression of values which run counter to a medical indication. Timely and appropriate physician-patient communication on treatment options and patient self-determination is in practice often an important prerequisite for achieving this goal.

The above-mentioned empirical studies and practical experience with psychiatric advance directives in the US show that shared decision-making between patient and psychiatrist on treatment matters can be achieved in clinical practice. Here, the instrument of the advance directive should be understood as an integral part of the joint process of communication and decision-making between patient and physician within the framework of a longer-term treatment plan. Furthermore, the frequency and quality of advance directives can be increased through targeted educational measures in the psychiatric hospital. This should not be forgotten in the debate on advance directives, which often focuses on the few cases where conflicts arise (Vollmann 2010; Zeug 2010). For, particularly in the treatment of mental disorders, the patient's individual personality, lifestyle and values play a key role. The utilization of advance directives in psychiatry presents an opportunity to improve psychiatric care through timely communication with patients and ascertainment of their self-determined wishes.

References

Appelbaum, Paul S. 1991. Advance directives for psychiatric treatment. *Hospital & Community Psychiatry* 42: 983–984.

Appelbaum, Paul S. 2004. Psychiatric advance directives and the treatment of committed patients. *Psychiatric Services* 55: 751–752, 763.

Appelbaum, Paul S., S.A. Mirkin, and A.L. Bateman. 1981. Empirical assessment of competency to consent to psychiatric hospitalization. *The American Journal of Psychiatry* 138: 1170–1176.

Appelbaum, Paul S., Thomas Grisso, Ellen Frank, Sandra O'Donnell, and David J. Kupfer. 1999. Competence of depressed patients for consent to research. *The American Journal of Psychiatry* 156: 1380–1384.

Bauer, Armin, and Jochen Vollmann. 2002. Einwilligungsfähigkeit bei psychisch Kranken Eine Übersicht empirischer Untersuchungen. *Der Nervenarzt* 73: 1031–1038.

Bean, G., S. Nishisato, N.A. Rector, and G. Glancy. 1994. The psychometric properties of the Competency Interview Schedule. *Canadian Journal of Psychiatry* 39: 368–376.

Beauchamp, Tom L., and James F. Childress. 1994. *Principles of biomedical ethics.* New York: Oxford University Press.

Breden, Torsten M., and Jochen Vollmann. 2004. The cognitive based approach of capacity assessment in psychiatry: A philosophical critique of the MacCAT-T. *Health Care Analysis* 12(4): 273–283; discussion 265–272.

Burchardi, Nicole, Oliver Rauprich, and Jochen Vollmann. 2005a. Patientenverfügungen in der hausärztlichen Betreuung von Patienten am Lebensende. *Palliativmedizin* 6(02): 65–69.

Burchardi, Nicole, Oliver Rauprich, Martin Hecht, Marcus Beck, and Jochen Vollmann. 2005b. Discussing living wills. A qualitative study of a German sample of neurologists and ALS patients. *Journal of the Neurological Sciences* 237(1–2): 67–74.

Carpenter, William T., James M. Gold, Adrienne C. Lahti, Caleb A. Queern, Robert R. Conley, Robert J. Bartko, Jeffrey Kovnick, and Paul S. Appelbaum. 2000. Decisional capacity for informed consent in schizophrenia research. *Archives of General Psychiatry* 57: 533–538.

Charland, Louis C. 1998. Appreciation and emotion: Theoretical reflections on the MacArthur Treatment Competence Study. *Kennedy Institute of Ethics Journal* 8: 359–376.

Deutsche Gesellschaft für Psychiatrie, Psychotherapie und Nervenheilkunde. 2009. *Für Patientenautonomie und ärztliche Fürsorge in der Psychiatrie. DGPPN: Stellungnahme zur Anhörung im Deutschen Bundestag über Patientenrechte und Patientenverfügungen. 3/03.03.2009.* http://www.dgppn.de/fileadmin/user_upload/_medien/download/pdf/pressemitteilungen/ 2009/pm-2009-03-patientenautonomie.pdf. Accessed 25 Nov 2010.

Deutsche Gesellschaft für Psychiatrie, Psychotherapie und Nervenheilkunde. 2010. *Auswirkungen des Betreuungsrechtsänderungsgesetzes (Patientenverfügungsgesetz) auf die medizinische Versorgung psychisch Kranker: Rechtsgutachten und Stellungnahme der DGPPN. 3 / 15.04.2010.* http://www.dgppn.de/fileadmin/user_upload/_medien/download/pdf/stellungnahmen /2010/stn-2010-04-15-patientenverfuegung.pdf. Accessed 25 Nov 2010.

Elbogen, Eric B., Jeffrey W. Swanson, Paul S. Appelbaum, Marvin S. Swartz, Joelle Ferron, Richard A. Van Dorn, and H. Ryan Wagner. 2007. Competence to complete psychiatric advance directives: Effects of facilitated decision making. *Law and Human Behavior* 31(3): 275–289.

Elliot, Carl. 1997. Caring about risks: Are severely depressed patients competent to consent to research? *Archives of General Psychiatry* 54: 113–116.

Emanuel, Linda L., Ezekiel J. Emanuel, John D. Stoeckle, Lacinda R. Hummel, and Michael J. Barry. 1994. Advance directives. Stability of patients' treatment choices. *Archives of Internal Medicine* 154(2): 209–217.

Finzen, Asmus. 2009. Ende der Zwangspsychiatrie? Das neue Gesetz zur Patientenverfügung erweitert den Vorsorgespielraum psychisch Kranker. *Psychosoziale Umschau* 04: 26.

Fritze, Jürgen, and Hans-Martin Sass. 2003. Patientenrechte, Patientenverfügung, Vorsorgevollmacht. *Der Nervenarzt* 74(7): 629–631.

Grisso, Thomas, and Paul S. Appelbaum. 1995. Comparison of standards for assessing patients' capacities to make treatment decisions. *The American Journal of Psychiatry* 152: 1033–1037.

Grisso, Thomas, and Paul S. Appelbaum. 1998. *Assessing competence to consent to treatment: A guide for physicians and other health professionals.* New York: Oxford University Press.

Grisso, Thomas, Paul S. Appelbaum, and Carolyn Hill-Fotouhi. 1997. The MacCAT-T: A clinical tool to assess patients' capacities to make treatment decisions. *Psychiatric Services* 48: 1415–1419.

Grossman, L., and F. Summers. 1980. A study of the capacity of schizophrenic patients to give informed consent. *Hospital & Community Psychiatry* 31: 205–206.

Hansen, H.C., R. Drews, and P.W. Gaidzik. 2008. Zwischen Patientenautonomie und ärztlicher Garantenstellung. Die Frage der Einwilligung von Patienten mit Bewusstseinsstörungen. *Der Nervenarzt* 79(6): 706–715.

Haupt, M., H. Seeber, and M. Janner. 1999. Patientenverfügungen und Bevollmächtigungen in gesundheitlichen Angelegenheiten älterer psychisch kranker Menschen. *Der Nervenarzt* 70(3): 256–261.

Helmchen, Hanfried. 1998. Research with patients incompetent to give informed consent. *Current Opinion in Psychiatry* 11: 295–297.

Helmchen, Hanfried, and Hans Lauter. 1995. *Dürfen Ärzte mit Demenzkranken forschen? Analyse des Problemfeldes Forschungsbedarf und Einwilligungsproblematik.* Stuttgart: Thieme.

Hoffman, B.F., and J. Srinivasan. 1992. A study of competence to consent to treatment in a psychiatric hospital. *Canadian Journal of Psychiatry* 37: 179–182.

Honberg, Ronald S. 2010. *Advance directives.* http://www.nami.org/Content/ContentGroups/Legal/Advance_Directives.htm. Accessed 16 June 2010.

Janofsky, Jeffrey S., Richard J. McCarthy, and Marshal F. Folstein. 1992. The Hopkins competency assessment test: A brief method for evaluating patients' capacity to give informed consent. *Hospital & Community Psychiatry* 43: 132–136.

Kim, Scott Y.H., Eric D. Caine, Glenn W. Currier, Adrian Leibovici, and J. Michael Ryan. 2001. Assessing the competence of persons with Alzheimer's disease in providing informed consent for participation in research. *The American Journal of Psychiatry* 158: 712–717.

Kitamura, Fusako, Atsuko Tomoda, Kazumi Tsukada, Makoto Tanaka, Ikuko Kawakami, Shuuichi Mishima, and Toshinori Kitamura. 1998. Method for assessment of competency to consent in the mentally ill: Rationale, development, and comparison with the medically ill. *International Journal of Law and Psychiatry* 21: 223–244.

Koch, Hans-Georg, Stella Reiter-Theil, and Hanfried Helmchen. 1996. *Informed consent in psychiatry. European perspectives of ethics, law and clinical practise.* Baden-Baden: Nomos.

Lang-Welzenbach, Marga, Peter A. Fasching, and Jochen Vollmann. 2005. Patientenverfügungen und Therapieentscheidungen in der gynäkologischen Onkologie – Qualitative Interviews mit Patientinnen, Ärzten und Pflegepersonal. *Geburtshilfe und Frauenheilkunde* 65(05): 494–499.

Lauter, Hans, and Hanfried Helmchen. 2006. Vorausverfügter Behandlungsverzicht bei Verlust der Selbstbestimmtbarkeit infolge persistierender Hirnerkrankungen. *Der Nervenarzt* 77(9): 1031–1039.

Lipp, Volker. 2009. *Handbuch der Vorsorgeverfügungen: Vorsorgevollmacht – Patientenverfügung – Betreuungsverfügung.* München: Vahler.

Mahler, John, and Samuel Perry. 1988. Assessing competency in the physically ill: Guidelines for psychiatric consultants. *Hospital & Community Psychiatry* 39: 856–861.

Markson, L.J., D.C. Kern, G.J. Annas, and L.H. Glantz. 1994. Physician assessment of patient competence. *Journal of the American Geriatrics Society* 42: 1074–1080.

Marson, Daniel C., Kellie K. Ingram, Heather A. Cody, and Lindy E. Harrell. 1995. Assessing the competency of patients with Alzheimer's disease under different legal standards. A prototype instrument. *Archives of Neurology* 52: 949–954.

Marson, Daniel C., Anjan Chatterjee, Kellie K. Ingram, and Lindy E. Harrell. 1996. Toward a neurologic model of competency: Cognitive predictors of capacity to consent in Alzheimer's disease using three different legal standards. *Neurology* 46: 666–672.

Marson, D.C., B. McInturff, L. Hawkins, A. Bartolucci, and L.E. Harrell. 1997. Consistency of physician judgements of capacity to consent in mild Alzheimer's disease. *Journal of the American Geriatrics Society* 45: 453–457.

McKinnon, Karen, Karen Cournos, and Barbara Stanley. 1989. Rivers in practice: Clinicians' assessment of patients decision-making capacity. *Hospital & Community Psychiatry* 40: 1159–1162.

National Resource Center on Psychiatric Advance Directives. 2010. *Homepage.* http://www.nrc-pad.org. Accessed 16 June 2010.

Nationaler Ethikrat. 2005. *Patientenverfügung.* Berlin: Nationaler Ethikrat.

Olzen, Dirk, Eylem Kaya, Angela Metzmacher, and Noëlly Zink. 2009. *Die Auswirkungen des Betreuungsrechtsänderungsgesetzes (Patienteverfügungsgesetz) auf die medizinische Versorgung psychisch Kranker. Gutachten für die Deutsche Gesellschaft für Psychiatrie.* Berlin: Psychotherapie und Nervenheilkunde.

Roth, Loren H. 1983. Is it best to obtain informed consent from schizophrenic patients about the possible risk of drug treatment, for example, tardive dyskinesia, before initiating treatment or at a later date? *Journal of Clinical Psychopharmacology* 3: 207–208.

Rutman, D., and M. Silberfeld. 1997. A preliminary report on the discrepancy between clinical and test evaluation of competence. *Canadian Journal of Psychiatry* 37: 634–639.

Sass, Hans-Martin, and Rita Kielstein. 2001. *Patientenverfügung und Betreuungsvollmacht*. Münster: Lit Verlag.

Srebnik, Debra S., Lindsay T. Rutherford, Tracy Peto, Joan Russo, Ellen Zick, Craig Jaffe, and Paul Holtzheimer. 2005. The content and clinical utility of psychiatric advance directives. *Psychiatric Services* 56: 592–598.

Swanson, Jeffrey W., S. Van McCrary, Marvin S. Swartz, Eric B. Elbogen, and Richard A. Van Dorn. 2006a. Superseding psychiatric advance directives: Ethical and legal considerations. *The Journal of the American Academy of Psychiatry and the Law* 34: 385–394.

Swanson, Jeffrey W., Marvin S. Swartz, Eric B. Elbogen, Richard A. Van Dorn, H. Joelle Ferron, Ryan Wagner, Barbara J. McCauley, and Mimi Kim. 2006b. Facilitated psychiatric advance directives: A randomized trial of an intervention to foster advance treatment planning among persons with severe mental illness. *The American Journal of Psychiatry* 163: 1943–1951.

Swartz, Marvin S., J.W. Swanson, R.A. Van Dorn, Eric E. Elbogen, and Martha Shumway. 2006. Patient preferences for psychiatric advance directives. *International Journal of Forensic Mental Health* 5: 67–81.

Vollmann, Jochen. 2000a. Chancen und Risiken von Patientenverfügungen bei dementiellen Störungen. *Zeitschrift für Gerontopsychologie und -psychiatrie* 13(1): 38–50.

Vollmann, Jochen. 2000b. *Aufklärung und Einwilligung in der Psychiatrie. Ein Beitrag zur Ethik in der Medizin*. Darmstadt: Steinkopff.

Vollmann, Jochen. 2000c. Einwilligungsfähigkeit als relationales Modell. Klinische Praxis und medizinethische analyse. *Der Nervenarzt* 71(9): 709–714.

Vollmann, Jochen. 2008. *Patientenselbstbestimmung und Selbstbestimmungsfähigkeit. Beiträge zur Klinischen Ethik*. Stuttgart: Kohlhammer.

Vollmann, Jochen. 2010. Leserbrief zu Kartin Zeug "Nicht gegen meinen Willen" (ZEIT Nr. 37–2010). *ZEIT* 65(39): 95.

Vollmann, Jochen, and Irene Knöchel-Schiffer. 1999. Patientenverfügungen in der klinischen Praxis. *Medizinische Klinik* 94(7): 398–405.

Vollmann, Jochen, and Michael Pfaff. 2003. Patientenverfügungen: Theoretische Konzeption und praktische Bedeutung in den USA. *Deutsche Medizinische Wochenschrift* 128(27): 1494–1497.

Vollmann, Jochen, Armin Bauer, Heidi Danker-Hopfe, and Hanfried Helmchen. 2003. Competence of mentally ill patients: A comparative empirical study. *Psychological Medicine* 33(8): 1463–1471.

Vollmann, Jochen, Klaus-Peter Kühl, Amely Tilmann, Heinz-Dieter Hartung, and Hanfried Helmchen. 2004. Einwilligungsfähigkeit und neuropsychologische Einschränkungen bei dementen Patienten. *Der Nervenarzt* 75(1): 29–35.

Welie, Sander P.K. 2001. Criteria for patient decision making (in)competence: A review of and commentary on some empirical approaches. *Medicine Health Care and Philosophy* 4: 139–151.

Wilder, Christine M., Eric B. Elbogen, Marvin S. Swartz, Jeffrey W. Swanson, and Richard A. Van Dorn. 2007. Effect of patients' reasons for refusing treatment on implementing psychiatric advance directives. *Psychiatric Services* 58: 1348–1350.

Wittink, Marsha N., Knashawn H. Morales, Lucy A. Meoni, Daniel E. Ford, Nae-Yuh Wang, Michael J. Klag, and Joseph J. Gallo. 2008. Stability of preferences for end-of-life treatment after 3 years of follow-up: The Johns Hopkins Precursors Study. *Archives of Internal Medicine* 168(19): 2125–2130.

Zeug, Katrin. 2010. Nicht gegen meinen Willen. *ZEIT* 37–2010: 38.

Part II

Defining the Scope of Advance Directives

Chapter 4
On the Scope and Limits of Advance Directives and Prospective Autonomy

Robert S. Olick

4.1 Introduction

Most patients near the end of life die after a decision is taken to withhold or withdraw life-sustaining interventions, such as cardiopulmonary resuscitation (CPR), a ventilator or a feeding tube. It has been estimated that in the US more than one million hospital patient deaths each year (70 % of all hospital deaths) occur after a decision to forgo life-sustaining interventions (Prendergast 2000). When the ravages of illness, disease or injury have stolen the patient's ability to decide for oneself, others must make these difficult decisions on behalf of incompetent loved ones. The ethical and legal consensus across the US holds that, when called upon to bear the burdens of decision, families and healthcare providers should seek, first and foremost, to determine what the patient would choose for him- or herself, and should also act in the patient's best interests. Every day in hospitals across the country, families, friends, physicians, social workers and others face the challenge of constructing a narrative of the patient's wishes, drawing on their understandings of the kind of person the patient has been over a lifetime, the patient's past statements, personal stories and experiences, and other information about the patient. In order to make patients' wishes count when we cannot speak for ourselves, and to ease the psychosocial and emotional burdens on family members, the widely adopted public policy response has been to encourage and empower individuals to write advance directives for healthcare to direct and control treatment decisions in the event of future decisional incapacity.

Focusing on the US experience, this chapter discusses key features of advance directives, and the scope and limits of their authority, from both ethical and legal perspectives. Some common ethical, legal and practical challenges for honouring directives are examined—in particular, dilemmas where ethics and law may not

R.S. Olick, JD, Ph.D. (✉)
Center for Bioethics and Humanities, SUNY Upstate Medical University,
Syracuse, NY, USA
e-mail: olickr@upstate.edu

P. Lack et al. (eds.), *Advance Directives*, International Library of Ethics,
Law, and the New Medicine 54, DOI 10.1007/978-94-007-7377-6_4,
© Springer Science+Business Media Dordrecht 2014

agree on the duty to honour patients' advance directives. Whether advance directives should be understood as binding documents is also considered.

There are generally three types of advance directives. A *proxy directive* (also known as a "durable power of attorney for healthcare") designates a trusted family member, friend or religious adviser to make healthcare decisions on the patient's behalf. The *living will* (also known as an "instruction directive") states with some specificity the person's wishes and instructions for care. The third approach is to designate a healthcare proxy and provide the proxy with further written instructions for future treatment and care, often called a *combined directive*. Most state statutes recognize both proxies and living wills and allow for combined directives; three US states recognize by statute only the healthcare proxy (American Bar Association 2009). The legal and policy landscape across Europe and other countries similarly recognizes both proxy directives and living wills as instruments for planning ahead for end-of-life decisions (Brauer et al. 2008), though looking to patients' previously expressed wishes as a basis for decision has been unusual in most European countries until recently (Andorno et al. 2009).

This chapter is primarily concerned with the designation of a healthcare proxy, the preferred and most widely used form of advance directive in the US. Among the reasons for the prevalence of the proxy directive are the fact that it is simple to use and permits the proxy to engage in an informed-consent dialogue with the physician and respond prudently to the patient's current circumstances and treatment options. Healthcare proxies avoid many (not necessarily all) of the well-documented problems that arise with interpretation of living wills, which too often prove ambiguous and unhelpful because they were written years ago and may not adequately anticipate the patient's current condition and treatment options.

4.2   Background and Legal Context

In the US, all 50 states and the District of Columbia recognize the legal right of competent adults to write advance directives to direct and control healthcare decisions near the end of life, at a time of future decisional incapacity. Advance directive laws have been strongly influenced by American case law, where the right to refuse life-sustaining treatment was first established. In the seminal case of Karen Ann Quinlan (*In re Quinlan* 1976), Joseph Quinlan was appointed legal guardian for his 21-year-old daughter and was granted permission to request removal of the respirator sustaining her life in a persistent vegetative state (PVS). This is a condition in which all cognitive functions of the brain have been lost, resulting in complete unawareness of self and the environment. PVS patients retain some of the brainstem functions that regulate autonomic activities of the body, such as breathing, but when properly diagnosed, there is virtually no hope of recovery to a cognitive, sapient state (Multi-Society Task Force on PVS 1994). The New Jersey Supreme Court's opinion was the first to recognize a constitutional right to refuse unwanted bodily interventions, including life-sustaining treatment, and to hold that when patients are unable to exercise that right, family members may forgo life support on behalf of incompetent loved ones, on the basis of the patient's wishes and best interests (often referred to as "substituted judgment"). In the ensuing two decades, states across the country had their own much-publicized cases requiring judicial resolution. Though not bound to follow *Quinlan*, courts in other states consistently found the *Quinlan* opinion's reasoning persuasive and most often reached the same essential conclusions. Over the next 15 years, a judicial consensus emerged supporting patients' rights and the authority of family members to make end-of-life decisions for incompetent loved ones. Some courts ground these rights in the federal or state constitution, others look to the common law right of self-determination. A number of these cases, like *Quinlan*, involved PVS patients; others involved patients who were terminally ill (Cantor 1993; Olick 2001). The US Supreme Court's decision in the case of Nancy Beth Cruzan (1990), a young woman in PVS, reaffirmed the legal-ethical consensus. *Cruzan* also held that states may establish their own rules for end-of-life decisions, provided they do not unduly infringe upon patients' constitutionally protected rights to control their own healthcare (*Cruzan* 1990).

In the immediate aftermath of *Quinlan*, California enacted the first advance directives law, the California "Natural Death Act" (1976). In the ensuing years extending into the early 1990s, state after state responded to this problem of reconstructing a reliable account of the patient's wishes, to court decisions, and to the clarion call for expansion of autonomy-based rights of patients and families, by enacting advance directive laws. The oft-recited legal principle grounding advance directive laws is that incompetent patients have the same rights of self-determination as competent patients. Only the means for exercising these important rights should differ. To effectuate control over the dying process, advance directive laws establish the right to put one's wishes for future care in writing and impose obligations on physicians, hospitals, families and others to honour the patient's wishes. Advance directives are typically used to appoint a healthcare proxy and to direct withholding or withdrawal of life-sustaining treatment when irreversible disease or injury brings a severely diminished quality of life, marked by incapacity, loss of control, dependence on medical interventions, pain and suffering. Recognizing that for some individuals longevity itself is valued despite severely impaired quality of life, most states also allow use of directives to request continued life support.

The deeper grounding of advance directives resides in the ethical principle of prospective autonomy. This principle recognizes that future-oriented actions are integral to the moral life of autonomous persons. To take actions and decisions now that affect one's future, or the future of family, colleagues, co-workers and friends, to commit to personal projects and goals—in short, to think and plan ahead and make an investment in the future—is an essential feature of living the moral life. Buying insurance, seeing the doctor regularly, having health-conscious diet and habits, and making important decisions about medications or surgery all involve a view to shaping and promoting our interests in good health, both for its own sake and for the critical instrumental importance of health and well-being in the pursuit of projects, commitments and goals that matter to us. In contrast to contemporaneous autonomy, the concept of prospective autonomy holds that it is an expression of our moral agency to make plans and take actions now that are intended to control our healthcare in the future, including when we can no longer make decisions for ourselves. When we commit to writing a personal plan for control of the dying process, to guide the course of care and treatment in the face of future incapacity to make contemporaneous decisions, this is an exercise of prospective autonomy. Taking charge of the dying process in this way is a defining feature of what is meant by "the pursuit of death with dignity". Directing one's personally selected healthcare proxy to carry out one's wishes extends autonomy beyond one's ability to make contemporaneous informed decisions and asserts control and dignity in the dying process. Moreover, the importance of future-oriented plans and commitments for how we die (or the legacy we leave our families) survives loss of capacity to appreciate whether those plans and commitments are respected. It still matters whether our wishes, values and decisions near the end of life are honoured or disregarded, even if we can no longer know what decisions are taken by our proxy, family and physician (Olick 2001).

4.2.1 Advance Directives Versus Physician Orders

Advance directives are sometimes confused with do-not-resuscitate (DNR) orders and with physicians' orders for life-sustaining treatment (POLST). One source of this confusion is that although DNRs and POLST are designed primarily to document and give effect to contemporaneous treatment decisions, they can also be used to provide direction for treatment decisions taking effect in the near future (thus, in advance), if patients' decisional capacity is lost. The key distinction is that, as the terms imply, both DNR and POLST are physician orders. In addition, both are intended to implement the patient's wishes.

DNR orders direct that resuscitation efforts be withheld should the patient suffer a cardiac arrest. They are for the most part contemporaneous orders, typically for hospitalized patients (out-of-hospital DNRs are also recognized), and are sometimes time-limited—that is, the DNR order may need to be revisited and renewed periodically. A DNR/DNI (do-not-intubate) order adds that in the event of cardiorespiratory distress the patient is not to be intubated. By contrast, advance directives specify the patient's wishes in anticipation of future ill health, are expressly designed to encompass a wide range of choices (not limited to DNR) and are personal documents with no requirement that a physician be involved in the crafting process (though communicating with one's physician is recommended); they remain valid indefinitely unless changed or revoked by the patient. When the proxy refuses resuscitation in accordance with the patient's wishes, the DNR order implements this decision.

POLST has emerged as a recent complementary mechanism for making patient and family wishes count at the bedside. The POLST form records in a single, multi-page document all physician orders for end-of-life care, including DNR/DNI, feeding tubes, dialysis and whether the patient should have comfort measures to ease the dying process or aggressive interventions to prolong life. In this way, it combines features of both advance directives and DNR orders. But like DNR orders, POLST is primarily designed to document orders for the hospitalized patient's condition and treatment options now and in the near future, based on patient and proxy (or other surrogate) decisions (Hickman et al. 2008; Sabatino 2010). POLST is sometimes used to implement advance directives and proxy decisions, but these documents are not legitimated or governed by the advance directive laws summarized here.

4.2.2 Synopsis

The balance of this chapter describes in more detail many common features of advance directives and the supporting laws, with emphasis on the scope and authority of advance directives at the bedside. The next section discusses the determination of patient incapacity that triggers the role and authority of advance directives. I then address the rights and duties of healthcare proxies, physicians and other healthcare providers. Next, the two most significant differences among state laws that shape and limit patient and proxy rights and the authority of advance directives are explained. The first concerns forgoing of life support for patients who are neither terminally ill nor permanently unconscious (the medical-condition limitation); the second, the rules for forgoing of feeding tubes (the feeding-tube limitation). Also discussed is a third and increasingly common dilemma on which the law is often silent—whether a proxy's insistence on continued treatment can be overridden on grounds of "medical futility". Patient care dilemmas involving any of these three scenarios can create conflict between ethics and law, and can put those responsible for the patient's care in the difficult position of asking, "Although it's ethical, is it legal?" Finally, this chapter addresses in what sense advance directives are or are not *binding*, and goes on to discuss the problem of the rebel proxy who fails to fulfil their fiduciary duty to honour the patient's wishes. Because my focus is on bedside decisions, I assume for the sake of discussion a properly executed proxy directive and do not set forth the formal requirements (such as signing and witnessing) for writing directives or their practical limitations.

My focus here is on the use of advance directives under state law, but a further feature of the legal landscape should be noted. The Patient Self-Determination Act (PSDA) is a federal law, applicable across the country. This procedural law imposes obligations on hospitals and other healthcare facilities to ask patients and families if the patient has a directive, to document patients' "advance directive status", and to provide information about patients' rights and advance care planning. The PSDA is silent on matters of substantive rights; it defers to the states as the source of substantive rights and rules for end-of-life decisions (Ulrich 1999). There is no national law in the US that establishes uniform rights and duties of patients, proxies,

physicians and hospitals. Because law and practice can differ from one state to the
next, it is important for patients, families and practitioners alike to be familiar with
the law of the state in which they live, and receive and provide healthcare.

4.3   Scope, Authority and Limitations of Advance Directives

4.3.1   Decisional Incapacity: A Triggering Condition

Competent patients have the right to control their own healthcare; all adults over the
age of 18 are presumed competent (to have decisional capacity). Because advance
directives are intended to take effect only when the patient lacks capacity, the law
typically prescribes a process for assessing capacity when the patient's ability to
make informed decisions is in question. Initially, the attending physician bears this
responsibility, and is to evaluate and document in the medical record the nature,
extent, cause and likely duration of the patient's incapacity. A second, confirming
opinion is often required, especially if the patient has a history of developmental
or intellectual disability (New York Health Care Proxy Law 2007). If the patient
lacks decisional capacity, the locus of authority to make treatment decisions shifts to
the proxy.

Though not often expressly stated in law, it is common clinical practice to
employ a decision-specific approach to capacity assessments. Patients may be
able to make one sort of decision, but not another; for example, the patient may
be able to choose a spouse or adult child to serve as healthcare proxy, but at the
same time cognitive deficits impair the ability to understand and reason about his or
her medical condition and the risks, benefits and burdens of refusing dialysis,
surgery or other recommended treatments. Also, some patients have fluctuating
capacity, meaning that they may be unable to make certain decisions today, but with
improving cognitive skills or reduced need for pain control can engage in decision
making the next day (Ganzini et al. 2005). A physician determination of incapacity
triggers the authority of the healthcare proxy, but for patients who are interactive,
with some capacity for reasoning, this should not categorically exclude them from
the decisional process. Patients retain the authority to decide when they regain
capacity, and respect for autonomy encompasses enhancing opportunities for
patients to make their own decisions. On this decision-specific approach, widely
endorsed in medical ethics, the power and right to decide may shift between patient
and proxy, depending on the patient's condition and the nature and complexity of
the decision to be taken. Once the proxy's role has been established, it is generally
advisable to continue to involve proxies in the decisional process even when
patients are able to and do make certain decisions for themselves. The proxy should
be considered both partner and (potential) decision maker in the care of the patient.

4.3.2 Rights and Duties of Healthcare Proxies

The rights, duties and responsibilities of healthcare proxies and healthcare providers shape bedside decisions involving advance directives and are at the heart of advance directive laws. Both law and ethics uniformly establish that the designated proxy's first obligation is to make decisions consonant with the patient's own wishes and values. Secondarily, the proxy should act in the patient's best interests. This fiduciary responsibility is recited in advance directive laws across the country and in the standard text of proxy forms. In practice, the decision-making process often involves consideration of the written document *and* other evidence of the patient's wishes (sometimes called subjective factors) as well as the patient's best interests (sometimes called objective factors), collectively forming a narrative of the patient's values, interests and intent to support a treatment decision. The patient's personally selected and trusted proxy is generally accorded substantial deference in the evaluation and interpretation of the written document, other information about the patient's wishes and values (for example, past statements made to family or friends), and how well this construction of the patient's wishes and interests fits his or her current medical condition and supports a treatment decision (for example, to refuse a respirator). In the event of significant conflict or inconsistency between the written document and reported past verbal statements, the written directive would presumptively take priority. As noted below, only in certain narrow circumstances would there be ethical ground to challenge the proxy's authority.

It is commonly stated that proxies may make any and all healthcare decisions the patient could make if competent, subject to any limitations set forth by the patient in the proxy document (such as that the proxy must consult with a sibling or is not to put the patient in a nursing home). Generally, proxies may make decisions to provide or forgo life-sustaining interventions (the animating purpose of advance directives); consent to hospice care, out-of-hospital DNR orders, or discharge to home or a nursing home; carry out the patient's intent to donate organs; request a consultation; choose a physician; and make other choices that belong to competent patients. Because authority to consent to healthcare also entails control over personal health information, proxies control access to otherwise confidential patient information.

Yet taken literally, the "any and all" formulation can be misleading. US law generally holds that in the exercise of the right of informed consent, competent patients may refuse any unwanted bodily intrusions, including any form of life-sustaining interventions, regardless of their medical condition. By contrast, advance directive laws sometimes impose significant limitations on the scope of permissible patient choice, and hence on the proxy's power. First and most important, with respect to their application to life-threatening medical conditions, advance directive laws uniformly apply to patients who are terminally ill or permanently unconscious, but only sometimes expressly include those facing progressive, irreversible but currently non-terminal diseases (such as earlier stages of cancer or Alzheimer's disease). Statutes often define a "terminal condition" to mean "death within a short time" or within approximately 1 year; clinical practice often mirrors hospice reimbursement guidelines—a prognosis of 6 months or less to live. Narrow definitions of terminal illness or condition can effectively exclude patients with chronic, progressive, but not currently life-threatening disease from the rights that belong to terminally ill patients, unless state law expressly encompasses those with non-terminal conditions who nonetheless face an inevitable decline. The medical-condition limitation in many advance directive laws is heavily influenced by the fact that precedent court cases most often involved end-of-life decisions for patients who were either terminally ill or in PVS. It also reflects a legislative (and political) balancing between societal interests in preserving life and patient and family rights to decide when enough is enough.

Second, several state laws impose restrictions on the right to refuse artificially provided fluids and nutrition (the feeding-tube limitation). For example, in a small number of states, the patient's directive must provide specific authorization or direction to forgo a feeding tube, intravenous fluids or like modalities, or there must be some reasonable evidence that this is consistent with the patient's wishes (American Bar Association 2009). If this legal requirement is not met, the presumptive approach is to insert or continue use of a feeding tube. The position that special rules are needed to withhold and withdraw feeding tubes and intravenous fluids rests on the view that artificially provided fluids and nutrition are akin to food and water, constituting basic human caring that holds cultural, religious and symbolic significance, and that feeding tubes must therefore be provided unless the higher standard of evidence of the patient's wishes is met (Lynn 1986). This view held significant currency in the 1980s, arguably the period of most intense debate on this issue, but is clearly a minority position today. It is a distinct departure from the judicial consensus that uniformly has made no distinction between the patient's right to refuse a feeding tube and the right to refuse other medical modalities such as a ventilator, dialysis or antibiotics (Meisel 1992).

The medical-condition and feeding-tube limitations can present ethical-legal dilemmas for patients, proxies, physicians and hospitals—situations at the bedside where it is proper to ask, "Although it's ethical, is it legal?" The nature of this tension between ethics and law is explored further below, with illustrative case examples. (Other notable legal limitations on patient choice and proxy authority include restrictive rules in some states on forgoing treatment during pregnancy. Secondly, patients have no right to physician-assisted suicide pursuant to an advance directive once they have lost capacity to make their own informed decisions. Both of these circumstances are, however, extremely rare in practice).

4.3.3 Rights and Duties of Physicians and Other Healthcare Providers

When we appoint family or friend as proxy, we expect that physicians, hospitals and others will respect the proxy's authority and will honour their decisions.

Advance directive laws uniformly impose obligations to respect advance directives and contain specific provisions that shape both duties and rights of physicians and healthcare facilities. To promote compliance, physicians, hospitals and others are commonly granted immunity from liability and from professional discipline so long as they act in good faith, in accordance with the patient's wishes and with accepted medical standards. In fact, in the more than 30 years since California enacted the first advance directive law, there have been very few lawsuits brought by grieving family members concerning removal of life support, and fewer still claiming a directive was ignored or overridden (Lynch et al. 2008). Under "reciprocity" provisions in force nationwide, documents written in another state are to be honoured, so long as they are validly executed under the law of the patient's home state or the state where care is delivered. Typical requirements are that the document be signed, dated and witnessed (American Bar Association 2009).

Physicians and hospitals should give proxies substantial deference in the interpretation and application of advance directives and other information about the patient's wishes. But healthcare providers have no obligation to comply with requests that are outside the bounds of the proxy's powers. Nor can physicians be compelled to engage in behaviour contrary to law. A clear illustration is that physician-assisted suicide is expressly proscribed in advance directive laws and is illegal in most US states. The practice is legal in four states—most notably Oregon, where it is permitted solely on the basis of informed consent given by a competent, terminally ill patient (Ganzini et al. 2001). Thus, such a request from a proxy could and should be refused by the physician, even if based on an express statement in the proxy document.

An important limitation on the duty to comply with the patient's/proxy's decision concerns rights of professional conscience. Building on well-established ethical-legal consensus, advance directive laws commonly permit physicians and other healthcare professionals to decline to participate in the forgoing of life support on the ground that this would violate sincerely held professional, personal or religious commitments and values. To illustrate, when such conflicts arise, they may be based on the physician's strong belief that forgoing treatment is not in the patient's medical best interests or that compliance is contrary to the standard of practice, or perhaps the physician's principled objection to withdrawal of feeding tubes. Though couched in terms of professional rights of conscience, the approach is to balance the rights and interests of both patients and professionals. Physicians (and others) asserting a conscientious objection bear responsibility to notify patient and proxy and to facilitate a transfer of care to another clinician. An appropriate transfer is, by definition, to another physician who does not have this same objection and is prepared to honour the patient's/proxy's choices. Pending appropriate transfer, care must continue to be provided and patients may not be abandoned, but there would be no duty to comply with the decision to which the physician objects. Similar rights of conscience have been extended to private healthcare institutions, most often those with religious affiliations (such as a Catholic nursing home). Where recognized by law, the exercise of institutional conscience must often meet requirements that the institution have written policies, must provide notice of its policies to patients and families prior to or at the time of admission, and will arrange for transfer of care should a conflict arise after admission (Cugliari and Miller 1994).

4.3.4   Ethical-Legal Dilemmas: Some Case Illustrations

To show how legal limitations on permissible patient and proxy choice sometimes conflict with and constrain the ethically justified treatment decision, consider Anna, a 71-year-old mother of two adult children, who has been living for several years with the progressive decline associated with Alzheimer's disease. She appointed her husband as healthcare proxy a year before her diagnosis, giving him authority to make treatment decisions in accordance with her wishes and best interests. In their long discussions over the past decade, Anna has been firm in her views. She does not want to have her life prolonged through the inevitable decline of long-term chronic illness if she can no longer interact meaningfully with her family. In her husband's judgment, that time has come. Anna has now been hospitalized for a life-threatening pneumonia. Acting as her proxy, her husband refuses antibiotics. However, the physician's view is that the pneumonia is easily cured and is not caused by Anna's underlying Alzheimer's disease. Believing it his duty to advocate for his patient's medical best interests, he insists that antibiotics must be provided. Since Anna is not at this time dependent on a respirator or feeding tube and her pneumonia can be cured, she is deemed not to have a terminal condition.

Taking into account the proxy appointment, her husband's statements about Anna's wishes, and their close relationship over many years, we can posit a strong argument that Anna would not want antibiotics to treat her pneumonia, and that her husband's decision should be respected. Indeed, many of us would agree that to be sustained in this condition with an inevitable downward path and diminishing quality of life would be an undignified existence, imposing undue burdens on ourselves and our families. But if this state's advance directive law expressly authorizes withholding of life-sustaining interventions only when the patient is terminally ill (or permanently unconscious), the physician and hospital may well refuse to comply on the ground that withholding antibiotics under these circumstances is not legally permitted. There is a strong argument that the hus-band-proxy's decision is ethical, but those responsible for complying with the decision have an argument that it is not legal to do so, or at least that the law does not require compliance. By contrast, many advance directive laws take a broader view of patient rights, and expressly authorize refusal of treatment when the patient has a "progressive illness that will be fatal and is in an advanced stage", or when the patient has a progressive, irreversible condition and the burdens of aggressive treatment outweigh the benefits (Olick 2012). Under these laws, the ethical-legal conflict is anticipated and resolved in favour of patient rights and proxy authority.

Where embodied in law, the feeding-tube limitation also can create an ethical-legal dilemma. Consider Joseph, a 59-year-old man with a history of lung cancer. Two days after admission for an oncology follow-up, he suffered a serious stroke that left him minimally responsive and unable to make decisions for himself. A year earlier, Joseph had designated his sister as healthcare proxy with power to "make any and all healthcare decisions for me, except to the extent I state otherwise", using the standard short form popular in his home state. The attending physician believes Joseph is terminally ill but could live another 6 months with a nasogastric feeding tube. Joseph's sister refuses the feeding tube, stating that this is not what her brother would want. She recounts several conversations in which her brother spoke of not wanting to be maintained on machines if he were dying, in pain and unable to communicate meaningfully with others. And she describes him as a very active person, who valued his freedom and abhorred dependence on others. But she cannot recall any specific conversation with her brother about feeding tubes, and there is no reference to artificially provided fluids and nutrition in the proxy directive.

Under some advance directive laws (again a minority), the physician and hospital may well object that, absent any reasonable evidence Joseph had specifically contemplated and formed an opinion about forgoing a feeding tube, there is no legal authority to do so, and further that there is a duty to insert and maintain the feeding tube. (For the occasional physician who holds a principled objection to withholding artificial feeding, believing it to be ordinary and obligatory care, professional conscience further buttresses the refusal to comply, with a correlative duty to transfer care.) Joseph's sister may well have a compelling ethical argument to refuse the feeding tube, but the feeding-tube limitation gives the physician and hospital legal ground to object. It creates an ethical-legal dilemma that limits patient rights and proxy authority. By contrast, on the (majority) rule where no such distinctions are drawn and there is no such limitation, the proxy's decision should be honoured without conflict, respecting Joseph's wishes.

A third scenario that puts ethics and law in tension arises when proxies and families insist on continued aggressive interventions that physicians consider "medically inappropriate" or "futile". Consider Paul, a 39-year-old man with a history of acute myelogenous leukaemia (AML). Multiple attempts at induction chemotherapy have been unsuccessful. His condition now being refractory to treatment, his wife (who he appointed as proxy 3 years ago) has consented to a palliative treatment plan. But she insists that Paul be resuscitated in the event of cardiac arrest. Social workers believe her insistence on CPR is based partly on reluctance to let go and partly on the desire to allow time for other family members to say goodbye. Paul's physicians believe CPR would be futile. They think it highly unlikely he would survive the attempt and implore the proxy to consent to a DNR order. Frustrated, they call for an ethics consultation in the hope that it will support such an order.

Here, it is the physicians who argue that to further limit life-sustaining interventions—to place a DNR order—is ethically justified. Many would agree this to be an ethically sound position, one that finds support in the American Medical Association (AMA) Code of Medical Ethics statement that when "further intervention to prolong the life of a patient becomes futile … [the intent of treatment] should not be to prolong the dying process without benefit to the patient or others with legitimate interests" (American Medical Association 2010). But law provides scant reason for the physician to "just say no". Advance directive laws typically state that there is no duty to provide treatment that is contrary to "accepted professional standards" or that is "not medically appropriate", but this language is widely considered to be pro forma and its application to futility cases is untested. Case law offers few examples and no consensus on the physician's asserted right to say no to life-sustaining treatment over the patient/proxy's objection. The common experience in practice is that ethical arguments of physicians and hospitals ultimately yield to legal rights of patients and families to insist on continued interventions to sustain life regardless of quality. "Futile" treatment deemed legally required though ethically inappropriate is provided.

4.3.5 Summary

To summarize, advance directives are grounded in the ethical principle of prospective autonomy, on which it is an essential feature of the moral life to take actions and make plans now that are intended to foster and secure a desired future. With respect to decisions near the end of life, advance directives exercise that moral agency by asserting control over the dying process in accordance with our own values and wishes. The legal framework and daily practice facilitate planning ahead for important end-of-life decisions, empower patients and proxies to make the patient's wishes count after capacity to make contemporaneous informed decisions has been lost, and obligate physicians, hospitals and others to honour the patient's wishes and the proxy's authority. In general, ethics, law and practice in the US take an expansive approach to patient rights and proxy power near the end of life. Commentators have proffered strong arguments that the medical-condition and feeding-tube limitations on patient/proxy choice in force under the law of a minority of states are unjustified (Cantor 1993), but—as the scenarios above illustrate—these limitations can constrain bedside decisions where ethics and law collide. The practical consequence is that sometimes physicians and hospitals feel compelled to refuse to respect the patient's/proxy's decision, asserting the absence of legal authority to withhold a feeding tube or to forgo life-sustaining treatment for a patient with a progressive, irreversible, but non-terminal condition; they thereby frustrate and override the patient's directive and deny the patient the right to control and shape a more dignified dying process. Medical futility scenarios present an opposite dilemma, where—in the absence of legal rules to support the physician's judgment—patient/proxy autonomy often (but not always) trumps the physician's ethical view that continued aggressive efforts to sustain life offer the patient no medical benefit. Of critical importance here, the resolution of ethical-legal conflicts at the bedside can hinge on one's understanding of the interface between ethics and law. For those who read such legal limitations literally and adhere to the myth that "anything the law does not expressly permit, it therefore prohibits" (Meisel 1991), such conflicts will likely be more common and patient rights more often dishonoured. On the other hand, physicians and other healthcare providers committed to supporting the patient and proxy in an ethically sound decision will often choose to ignore the law, embrace the understanding that though the law does not authorize, it does not expressly prohibit forgoing of treatment in these circumstances, or otherwise reconcile their professional judgment to respect the patient's/proxy's decision.

4.4   Are Advance Directives Binding?

4.4.1   Meaning of the Concept

Discussions of the scope and limits of advance directives sometimes pose the question, "Are advance directives binding?" In order to answer this question, it is important to distinguish different senses in which the term "binding" has been used. Sometimes families believe that advance directives must always be followed, in effect that "Whatever the proxy says goes." Some physicians believe proxy directives to be binding in this way as well; others have difficulty relinquishing control over treatment decisions and resist or show ambivalence towards the proxy's role. It should be evident from the foregoing that if what is meant by "binding" is that the literal terms of a directive must always be strictly followed, or that the proxy's choice must always be honoured whatever it is, then directives are not binding in this absolute sense. When ethics and law collide, we may hold that there is an ethical obligation to honour the patient's wishes and the proxy's authority. But it is difficult to maintain that physicians and hospitals are duty-bound to disregard the law and put themselves at risk for legal entanglement or professional discipline, even given that lawsuits involving advance directives are rare and risk of legal liability low (hence advance directives are also not binding in the sense that they are legally enforceable) (Lynch et al. 2008).

Proxy directives are, however, binding in a third and very important sense. A properly executed document imposes ethical and legal obligations on the proxy, family, physician and other healthcare providers to honour the patient's wishes, and those directions may be overridden only if there is strong justification for doing so. Rephrased, advance directives are *prima facie* binding. They are to be respected unless those who question the obligation to honour the directive and the proxy's decision establish strong justification to override it. A number of European countries embrace this position as well, though the Council of Europe's Biomedicine Convention (1997) adopts a weaker position, stating that patients' previously expressed wishes "shall be taken into account" (Andorno et al. 2009). As I have argued elsewhere (Olick 2001), ethical grounds to either remove the proxy or override a particular decision arise in *the case of the rebel proxy* who fails to fulfil his or her fiduciary duty to the patient. Putting aside the prior discussion of *legal* rules that limit the scope of proxy choice, the next section summarizes the position that, in rare cases, there is a strong *ethical* argument to override the proxy. This position is generally supported in law, though not articulated in this way.

4.4.2 The Case of the Rebel Proxy

As an exercise of prospective autonomy, the writing of advance directives entails anticipatory judgment. A personally selected spouse, partner, adult child, or friend designated as a healthcare proxy is entrusted with the role of faithfully assuming the responsibility to make decisions in accord with the patient's wishes and best interests. But on rare occasion that faith and trust is violated. Most scholarly discussion points to the proxy who acts from malicious, malevolent or self-interested motives, intending to impose undue suffering or to hasten death and the path to a tidy inheritance (sometimes called a "turncoat proxy"). The proxy who not only harbours such feelings and intentions but also seeks to act on them—perhaps offering a tortured account of the patient's wishes out of ulterior motive—should be stripped of the power to decide. One may also be a *rebel proxy* in a more benign and more common way, by failing to honour the patient's wishes despite good-faith efforts to do so. From time to time, a proxy is unable to shoulder the burdens of decision; clearly contravenes written instructions or family consensus about the patient's wishes because of an inability to let go; strains too hard to rely on a hopelessly ambiguous directive, contrary to the best interests of the patient; or simply is "dead wrong" in their understanding of the patient's wishes.

A familiar example in clinical practice involves apparent conflict between a living will and the proxy's decision. Consider the patient who chooses his spouse as proxy but fails to discuss the meaning of a living will authored several years before, or whose combined directive conjoins the proxy appointment with the more detailed instructions of a living will. Critics of living wills often contend that these instructions are vague, ambiguous and unhelpful as decisional tools (Fagerlin et al. 2002). Experience with living wills that may have been written years ago in times of good health shows that we can be poor forecasters of future illness, disease and disability and of what our medical needs and options will be. Further, living wills have often used vague and ambiguous language that refuses treatment when there is "no reasonable expectation of my recovery from physical or mental disability" or simply refuses "heroic measures" (Eisendrath and Jonsen 1983). Because living wills commonly set forth with some specificity interventions that are or are not wanted in particular medical circumstances, they give the appearance of clear guidance and direction, and often this is the case. But when the patient's actual medical circumstances and treatment options differ substantially from those previously contemplated, the specificity of a living will can create ambiguity and uncertainty. A fair reading of the patient's express contemplation of a different medical situation may be that the patient did not consider his or her current condition and options. Perhaps the patient did not want to be ventilator dependent but might accept a trial of ventilation. Or, analogous to the case of Anna, the living will clearly refuses ventilator and feeding tube, but is silent about antibiotics. When healthcare providers or even other family members take the literal written word (the living will) as the controlling statement of the patient's wishes, they sometimes question the authority of the proxy who would decide otherwise. Or, those responsible for the patient's care may conclude that the document fails to provide sufficient guidance about the patient's wishes, or that the document is hopelessly ambiguous when applied to the patient in the bed before them. They may therefore question the proxy's reliance on the content of the living will, and take the position that the proxy's decision cannot be accepted.

Another scenario can also be imagined. Future-oriented refusals of treatment are rooted in today's understanding of the nature of illness, disease and disability, and the potential and limitations of medicine to heal, restore function, control pain and relieve suffering. Occasionally, new developments in medicine not previously contemplated by the patient's directive will offer significant promise of substantial benefit for the patient. Consider, for example, the promise of ongoing research into the genetic basis of common diseases such as cancer and the quest for more effective pharmacogenetic interventions; perhaps a biomedical research break-through to reverse the course of Alzheimer's disease is on the horizon. When what medicine has to offer has changed substantially, when there has been a *radical change in circumstances* from those previously contemplated by the patient, there is good reason *not* to honour the patient's prior refusal of treatment. Of course, medicine is constantly changing. How radical, then, must the change in what medicine has to offer be to justify overriding the patient's refusal? It must be an intervention that promises to alleviate the very unwanted conditions that underlie the refusal of treatment—incapacity, pain, suffering, dependence on others, or other conditions material to the patient's view of an unacceptable quality of life. A "new" treatment that would merely prolong life a while more does not count as offering substantial benefit to the patient.

A radical change in circumstances presents a strong ethical argument for consenting to the intervention and not following a prior refusal of life-sustaining treatment. The proxy who chooses this course may justify this position on either of two grounds: first, that the patient would not have intended to refuse life-sustaining interventions had s/he known or anticipated this new and effective treatment; and second, that in this unusual case, the proxy's authority to decide in the patient's best interests trumps rigid adherence to the patient's prior wishes, which are now of uncertain meaning and application. Physicians and others ought to concur in this judgment and honour the proxy's decision. But suppose the proxy ignores or fails to take account of this radical change in circumstances. These very same arguments would support the efforts of physicians or other family members to insist on the new and beneficial treatment and to override the proxy's objection even where refusal of treatment is based on a reading of the patient's living will.

These are real, though rare, possibilities. Any of these scenarios may justifiably give pause and warrant closer scrutiny, inclining physicians and hospitals to challenge the proxy's authority. But only rarely will overriding the proxy ultimately be justified. Again, the proxy is due substantial deference and respect in his or her interpretation of the directive and of the patient's wishes. This includes insights into what the patient means by "no heroic measures", "no meaningful quality of life", or other such phrases. The proxy has been entrusted with exercising sound judgment to understand what such phrases mean to the patient, and to place them in context so as to clarify what the patient values and what the patient would find an unacceptable quality of life or an undignified dying process. Moreover, when evidence of the patient's wishes is lacking or hopelessly ambiguous, the proxy has residual authority to decide in the patient's best interests. When confronted with a rebel proxy, we may say that the patient chose unwisely. But absent extraordinary circumstances (a coerced proxy appointment), a valid proxy designation is nonetheless the patient's prospectively autonomous choice and is entitled to the strongest presumption of respect. Hence, healthcare professionals ought not simply to proceed with the course of treatment they think best for the patient and ignore or override the proxy; this would fail to take prospective autonomy and advance directives seriously. Rather, further process is in order, such as resort to an ethics consultation to first seek to resolve the dilemma, or to a court of law. To override the proxy may require formal judicial review in some states (New York Health Care Proxy Law 2007). It bears emphasis that in the case of the turncoat proxy of ill motive, removing the proxy's authority qua proxy would be an appropriate response. For the more familiar case of the rebel proxy, acting in good faith but not following the patient's wishes, it is appropriate to seek to challenge the particular decision but is often also proper to continue to involve the proxy in the patient's care and to look to the proxy to fulfil his or her fiduciary role with respect to other treatment and care decisions.

4.5   Conclusion

Since the New Jersey Supreme Court's decision in the case of Karen Ann Quinlan (1976), the right to refuse treatment has become firmly established in both ethics and law. Grounded in the ethical principle of prospective autonomy, legislation nationally has recognized the right to use advance directives to plan ahead to control treatment decisions in anticipation of future illness, disease and disability that takes away one's decisional capacity and prevents contemporaneous, informed choice. Law has played a prominent role in shaping the paradigm shift in society and medicine from a long history of "doctor knows best" to contemporary norms that put patient and proxy voice at the centre of decisions near the end of life. Strong ethical and legal support for advance directives does not mean, however, that prospective autonomy is unfettered—that any and all end-of-life decisions based on evidence of the patient's wishes and values must be honoured, or that anything the proxy decides must be done. Proxy directives are prima facie binding. Ethics and law accord substantial deference to proxy decisions and require those decisions to be honoured unless those who question this obligation establish strong justification for non-compliance. The exercise of prospective autonomy and proxy power encounters a number of ethical, legal and practical limitations. Most notably, law is not uniform across the US. A number of states limit the right to refuse treatment to conditions of terminal illness and permanent unconsciousness, and a handful impose special requirements on forgoing of feeding tubes. In those rare cases where a proxy fails to meet the fiduciary duty to decide in accord with the patient's wishes and best interests, it is justified to challenge and override the status or decisions of the rebel proxy.

Still, only about 20 % of US citizens write advance directives. This figure has not changed dramatically over time (Perkins 2007). Higher incidence of use has been reported among nursing home patients (Molloy et al. 2000), the elderly (AARP 2008), college graduates (Mueller et al. 2010) and people living with HIV/AIDS (Teno et al. 1990). A number of proposals have been made to increase the use and effectiveness of advance directives. These include reimbursing physicians for end-of-life discussions with patients (Fried and Drickamer 2010); making advance directive forms more widely available in languages other than English, removing the standard two-witness requirement, which sometimes disenfranchises the unbefriended elderly, and broadening eligibility rules for who may be chosen as proxy (Castillo et al. 2011); and making forms more readable and user-friendly (Mueller et al. 2010). All are worthy of pursuit but likely to achieve only modest gains. The psychosocial complexities of facing mortality and engaging the questions of how we die, and doing so in meaningful dialogue with family and physician, are intrinsic barriers to advance care planning. More often than not, family, friends, physicians and others are and will be called upon to decide for patients without the guidance and direction of advance directives.

References

American Association of Retired Persons. 2008. AARP bulletin poll "getting ready to go" executive summary. http://www.aarp.org/relationships/grief-loss/info-01-2008/getting_ready. html. Accessed 12 Oct 2012.
American Bar Association Commission on Law and Aging. 2009. Health care power of attorney and combined advance Directive Legislation. http://new.abanet.org/aging/Pages/HealthDecisions. aspx. Accessed 12 Oct 2012.
American Medical Association and Council on Ethical and Judicial Affairs. 2010. *Code of medical ethics: Current opinions with annotations, 2010–2011*. Chicago: American Medical Association.
Andorno, Roberto, Nikola Biller-Andorno, and Susanne Brauer. 2009. Advance health care directives: Towards a coordinated European policy? *European Journal of Health Law* 16(3): 207–227.
Brauer, Susanne, Nikola Biller-Andorno, and Roberto Andorno. 2008. *Country reports on advance directives*. www.ethik.uzh.ch/ibme/.../2008/ESF-**CountryReports**.pdf. Accessed 15 June 2012.

Cantor, Norman L. 1993. *Advance directives and the pursuit of death with dignity*. Bloomington: Indiana University Press.

Castillo, Lesley S., Brie A. Williams, Sarah M. Hooper, Charles P. Sabatino, Lois A. Weithorn, and Rebecca L. Sudore. 2011. Lost in translation: The unintended consequences of advance directive law on clinical care. *Annals of Internal Medicine* 154(2): 121–128.

Cruzan v. Director, Missouri Dept. of Health, 497 U.S. 261 (1990).

Cugliari, Anna Maria, and Tracy E. Miller. 1994. Moral and religious objections by hospitals to withholding and withdrawing life-sustaining treatment. *Journal of Community Health* 19(2): 87–100.

Eisendrath, Stuart J., and Albert R. Jonsen. 1983. The living will: Help or hindrance? *Journal of the American Medical Association* 249(15): 2054–2058.

Fagerlin, Angela, Peter H. Ditto, Nikki A. Hawkins, Carl E. Schneider, and William D. Smucker. 2002. The use of advance directives in end-of-life decision making. *American Behavioral Scientist* 46(2): 268–283.

Fried, Terri R., and Margaret Drickamer. 2010. Garnering support for advance care planning. *Journal of the American Medical Association* 303(3): 269–270.

Ganzini, Linda, Heidi D. Nelson, Melinda A. Lee, Dale F. Kraemer, Terri A. Schmidt, and Molly A. Delorit. 2001. Oregon physicians' attitudes about and experiences with end-of-life care since passage of the Oregon death with dignity act. *Journal of the American Medical Association* 285(18): 2363–2369.

Ganzini, Linda, Ladislav L. Volicer, Willam A. Nelson, Ellen Fox, and Arthur R. Derse. 2005. Ten myths about decision-making capacity. *JAMDA* 6: S100–S104.

Hickman, Susan E., Charles P. Sabatino, Alvin H. Moss, and Jessica Wehrle Nester. 2008. The POLST (Physician Orders for Life-Sustaining Treatment) paradigm to improve end-of-life care: Potential state legal barriers to implementation. *The Journal of Law, Medicine & Ethics* 36(1): 119–140.

In re Quinlan, 355 A.2d 647, *cert. denied sub nom. Garger v. New Jersey*, 429 U.S. 922 (1976).

Lynch, Holly F., Michele Mathes, and Nadia N. Sawicki. 2008. Compliance with advance directives: Wrongful living and tort law incentives. *Journal of Legal Medicine* 29: 133–178.

Lynn, Joanne (ed.). 1986. *By no extraordinary means: The choice to forgo life-sustaining food and water*. Bloomington: Indiana University Press.

Meisel, Alan. 1991. Legal myths about terminating life support. *Archives of Internal Medicine* 151(8): 1497–1502.

Meisel, Alan. 1992. The legal consensus about forgoing life-sustaining treatment: Its status and prospects. *Kennedy Institute of Ethics Journal* 2(4): 309–345.

Molloy, William D., Gordon H. Guyatt, Rosalie Russo, Ron Goeree, Bernie J. O'Brien, Michel Bedard, Andy Willan, Jan Watson, Christine Patterson, Christine Harrison, Tim Standish, David Strang, Peteris J. Darzins, Stephanie Smith, and Sacha Dubois. 2000. Systematic implementation of an advance directive program in nursing homes. *Journal of the American Medical Association* 283(11): 1437–1444.

Mueller, Luke A., Kevin I. Reid, and Paul S. Mueller. 2010. Readability of state-sponsored advance directive forms in the United States: A cross sectional study. *BMC Medical Ethics* 11(6): 1–6.

Multi-Society Task Force on PVS. 1994. Medical aspects of the persistent vegetative state (1). *The New England Journal of Medicine* 330(21): 1499–1508.

New York Health Care Proxy Law. *N.Y. Pub. Health Law* §2992 (West 2007).

Olick, Robert S. 2001. *Taking advance directives seriously: Prospective autonomy and decisions near the end of life*. Washington, DC: Georgetown University Press.

Olick, Robert S. 2012. Defining features of advance directives in law and clinical practice. *Chest* 141: 232–238.

Perkins, Henry S. 2007. Controlling death: The false promise of advance directives. *Annals of Internal Medicine* 147(1): 51–57.

Prendergast, Thomas J. 2000. Withholding or withdrawal of life-sustaining therapy. *Hospital Practice* 35(6): 91–102.

Sabatino, Charles P. 2010. The evolution of health care advance planning law and policy. *The Milbank Quarterly* 88(2): 211–239.

Teno, Joan, John Fleishman, Dan W. Brock, and Vincent Mor. 1990. The use of formal advance directives among patients with HIV-related disease. *Journal of General Internal Medicine* 5: 490–494.

Ulrich, Lawrence. 1999. *The patient self-determination act: Meeting the challenges in patient care*. Washington, DC: Georgetown University Press.

Chapter 5
Revocation of Advance Directives

Ralf J. Jox

5.1 Informed Consent and Revocation

It has been a cornerstone of informed consent that the person granting consent may also revoke or withdraw it at a later time. This principle is fully accepted for consent to both medical research and treatment.

The Declaration of Helsinki (World Medical Association 2008), the primary source of ethical guidance for research involving human subjects, states in Paragraph 24:

> The potential subject must be informed of the right to refuse to participate in the study or to withdraw consent to participate at any time without reprisal.

The separation of research and care is affirmed in Paragraph 34:

> The refusal of a patient to participate in a study or the patient's decision to withdraw from the study must never interfere with the patient-physician relationship.

The guarantee that research participants may freely withdraw from a study is considered one of the criteria that make research on human subjects ethical in the first place (Emanuel et al. 2000). Similarly, informed consent as a legal and ethical prerequisite for medical treatment entails the right of the patient to withdraw his or her consent at any time (Faden et al. 1986; Bernat 2001).

The philosophical foundation of this right is essentially the same as that of the right to refuse research participation or medical treatment from the outset. It is the individual autonomy of persons that gives them the right to be free from undesired intrusion into bodily integrity and privacy. There may also be additional foundations for this right, such as the aims of protecting the well-being of vulnerable persons or ensuring public trust in research and healthcare (Schaefer and

R.J. Jox, M.D., Ph.D. (✉)
Institute of Ethics, History and Theory of Medicine, Ludwig-Maximilians-University,
Munich, Germany
e-mail: ralf.jox@med.uni-muenchen.de

P. Lack et al. (eds.), *Advance Directives*, International Library of Ethics,
Law, and the New Medicine 54, DOI 10.1007/978-94-007-7377-6_5,
© Springer Science+Business Media Dordrecht 2014

Wertheimer 2010). As personal autonomy is usually regarded as entailing the right to waive autonomous decisions and restrict one's own scope of action, it is in principle conceivable that, when giving consent, someone may waive his or her right to subsequently withdraw it—as this might reduce drop-out rates in research studies and thus enhance their scientific validity. While some scholars hold that this option in fact protects personal autonomy (Edwards 2005), others maintain that the right to withdraw consent is inalienable (McConnell 2010).

Medical treatment—and to a lesser extent also research—is, however, also practised when persons have lost the capacity to give informed consent. The concept of decision-making capacity that is applied here is inherently normative. It is usually linked to the individual's communicative, deliberative and volitional capabilities: he or she must understand and retain the information relevant to the decision, weigh the arguments for and against a proposed form of treatment or research on the basis of individual preferences and values, derive an authentic and voluntary decision from this balance of arguments, and communicate this decision to others (Appelbaum 2007; Johnston and Liddle 2007; Legislation.gov.uk 2012). Importantly, this means that someone does not lack capacity merely because the decision he or she makes appears self-damaging, irrational or incomprehensible. For adults at least, capacity is seen as the default and incapacity has to be proven. It is usually assumed that incapacity is the result of some form of illness, impairment or disturbance of the mind or brain—be it temporary or permanent (although it is unclear whether this disturbance has to be established as a condition for incapacity or is simply a consequence of incapacity). The principle of respect for autonomy implies that every effort must be made to enhance an individual's decision-making capacity before denying him or her the status of full capacity.

Advance directives are an attempt to safeguard patient autonomy in the state of incapacity. In the broader sense, the term embraces both specific instructions about desired or undesired future care (instructional advance directives, also called living wills) and the appointment of a surrogate decision maker (proxy advance directives), commonly effected via the legal instrument of durable power of attorney (Rich 2004). In the stricter sense (the one used throughout this article), the term refers only to instructional advance directives. While these were originally developed for medical treatment, specifically at the end of life (Kutner 1969), they can also be used for research (Stocking et al. 2006). It is crucial to recognize that advance directives can be used both to grant and deny consent to future interventions in care or research. The former use is often found in psychiatric or research directives (Henderson et al. 2008; Srebnik et al. 2005; Muthappan et al. 2005), the latter more often in directives on end-of-life care (Silveira et al. 2010; Nicholas et al. 2011).

While the classical concept of informed consent rests on the situational, contemporaneous exercise of patient autonomy, advance directives are based on what has been called precedent or prospective autonomy (Davis 2002; Rich 1997). It is an open philosophical question whether the two forms of autonomy share the same moral authority or whether prospective is inferior to contemporaneous autonomy. Both rest on the conditions of decision-making capacity and voluntariness.

It seems, though, that the obligation to adequately inform the patient is more emphasized in classical informed consent than in anticipatory consent by advance directives (maybe because the latter are largely used for anticipatory refusal). As for informed consent (see above), the possibility of revoking one's consent or refusal is also a pivotal part of advance directives. In clinical practice, however, there are many situations where it is unclear whether a patient has revoked his or her directive or whether the preferences laid down in it still reflect the current preferences. This problem of practical significance has not yet been adequately addressed by bioethical scholars.

I will approach this problem in three steps: firstly, I will review the empirical evidence regarding the stability of patients' treatment preferences, as this has an impact on the issue of revocation of advance directives. Secondly, I will discuss various aspects of revocation, such as content, form, time and state of mind. Thirdly, I will describe, classify and discuss the ethical significance of common examples of patient behaviour that may be interpreted as expressions of "natural will" revoking an advance directive. Throughout, I will focus exclusively on directives for medical treatment, since research directives are not yet widely used and the main ethical problems arise in healthcare—particularly in paediatrics, geriatrics, psychiatry, critical care and palliative care.

5.2 Stability of Treatment Preferences

Why should we think about the stability of treatment preferences? The reason is that advance directives essentially document an individual's treatment preferences, and their revocation is tied to an alteration in these preferences—what else could motivate someone to change his or her directive? Of course, we assume that the patient is free to state his or her own preferences. If the patient is forced (as a result of duress, coercion, undue influence, deception or manipulation) to state preferences that are not authentic, then this is a different situation. In exceptional cases, the patient might be urged by a malicious relative or ruthless prospective heir to revoke a previously expressed refusal of, or consent to, treatment, depending on this person's emotional, financial or other interests. Yet we should not conflate use and abuse. These are certainly very rare situations. It is the responsibility of relatives, surrogates and healthcare professionals to be vigilant, identify such cases and take the appropriate (if necessary legal) action.

Even a patient who is free to state his or her own preferences might still have a poor memory and be mistaken about the content of an advance directive written long ago (e.g. if the document is not accessible). But as an advance directive encapsulates a major decision in life (like the decision whether to donate one's organs), it is highly unlikely that such a forgetful person will still have decision-making capacity. Rather, such forgetfulness concerning decisions of vital importance would be suggestive of mild dementia or other cognitive disorders. If a person of sound mind was mistaken about the content of an advance directive and erroneously thought it needed to be revoked, then that person's new preferences would probably be the same as the old ones, so this would make no practical difference.

If we rule out these two highly unrealistic situations, a change in an advance directive will probably reflect a change of mind. In fact, one of the arguments most frequently heard against advance directives is that, on the assumption that patients will often change their minds in the course of a severe illness, their directives will not reflect their true, current preferences. The stability of treatment preferences has therefore been studied empirically in numerous research projects. Unfortunately, there are no studies examining real-world treatment decisions in a prospective longitudinal design. Instead, researchers assessed preferences on the basis of hypothetical case vignettes, using questionnaires or specific scales at different points in time.

The (methodologically) best studies with the largest samples unanimously found that 70–85 % of study participants (mostly elderly subjects in an outpatient setting) have stable preferences for up to 2 years with regard to end-of-life treatment and, in particular, the use of life-sustaining measures such as cardiopulmonary resuscitation (CPR) (Danis et al. 1994; Krumholz et al. 1998; Carmel and Mutran 1999; Weissman et al. 1999; Bosshard et al. 2003; Ditto et al. 2003). Some authors calculated the inter-rater (Cohen's kappa) coefficient as a measure of stability of preferences and found values between 0.21 and 0.48, signifying a fair level of agreement which is quite good for longitudinal studies (Pruchno et al. 2008; Everhart and Pearlman 1990; Landis and Koch 1977; Emanuel et al. 1994).

Most studies also investigated which factors correlate with higher rates of stability or could even predict stability of preferences. Higher rates of stability were found in those individuals who were against life-sustaining treatment measures, as well as in those who had written advance directives (Danis et al. 1994; Carmel and Mutran 1999; Weissman et al. 1999; Ditto et al. 2003; Wittink et al. 2008). This seems plausible, as most authors of advance directives express their scepticism about life-sustaining treatment. From qualitative studies, we also know that people writing advance directives are often very self-determined, independent personalities with firmly held convictions (Inthorn 2008). Some studies found a correlation between stability of preferences and a higher level of education, certain ethnicities, or previous patient-physician communication on treatment preferences (Weissman et al. 1999; Emanuel et al. 1994; Ditto et al. 2003; Froman and Owen 2005). Another finding of these studies is that preferences are more stable for some clinical situations than for others: higher stability was reported for preferences regarding minor illnesses and very severe clinical scenarios such as a vegetative state; lower rates of stability were found for dementia (Weissman et al. 1999; Ditto et al. 2003).

Changes in treatment preferences were found to occur in the context of changing physical or mental health states, accidents or hospitalization (Danis et al. 1994; Weissman et al. 1999; Ditto et al. 2003; McParland et al. 2003; Schwartz et al. 2004). A study with a large sample of elderly individuals revealed that acceptance of life-sustaining treatment plummeted shortly after hospitalization but returned to baseline levels several months after hospitalization (Ditto et al. 2006). This observation—that changes in preferences might be temporary and only last for a short time during a critical health situation—is corroborated by the psychological insight that people have different preferences when they are in affectively "hot" as opposed to "cold" states, e.g. in pain versus a painless state (Loewenstein 2005). Those patients who changed their preferences over time tended to lose their interest in life-sustaining measures (Berger and Majerovitz 1998; Danis et al. 1994; Carmel and Mutran 1999; Ditto et al. 2003; Martin and Roberto 2006). There was, however, one exception: two studies on preferences for CPR—one in patients suffering from chronic heart disease, the other in patients with AIDS—reported that the number of patients who wanted CPR increased over time (Weissman et al. 1999; Krumholz et al. 1998).

To sum up, empirical studies consistently show that end-of-life treatment preferences are fairly stable, particularly if they tend to reject the administration of life-sustaining treatment. Hence, it is reasonable to assume that the preferences documented in an advance directive still represent the patient's authentic and essentially unchanged preferences. It would not be justifiable to infer that preferences have changed on the grounds that the document was written long ago, or that the patient has since experienced a marked decline in health. Even patients known to have a volatile, capricious personality will not necessarily have changed their minds regarding fundamental convictions expressed in an advance directive.

On the other hand, nobody should be surprised if patients' preferences have in fact changed during the course of their illness. Patients have a responsibility—and should be advised accordingly—to update their directives whenever their underlying preferences shift. Counsellors will always recommend that the documents should be regularly reviewed, e.g. once a year or every other year. In some countries, advance directives lose their validity after a defined time span: Austrian law requires citizens to renew legally binding directives every 5 years (Koertner et al. 2007; Schaden et al. 2010). Even in countries where the legal validity of advance directives is not dependent on the date of issue, physicians may become sceptical if a directive is very old and was issued in a totally different period of the patient's life.

Advance care planning programmes seek to ensure that directives are kept up to date and that any change in preferences is documented as soon as it occurs (Emanuel 2008). If they live up to this promise, they can relevantly enhance the authenticity and credibility of advance directives. Persons who have chosen to issue a directive will usually have a low threshold for changing it once they come to different views on these fundamental questions. A survey has shown that terminally ill patients attach more importance to their advance directive than healthy people (Jox et al. 2008), so it can be expected that they will also modify their directive when their preferences change. Nevertheless, it is conceivable that someone may have changed their mind but still hesitate to modify their directive, as they may not have found the right words or their thoughts may still be in flux. If such a patient suddenly loses decision-making capacity, relatives have to give credible testimony of the change in preferences and explain satisfactorily why the advance directive has not (yet) been modified. Many laws on advance directives have provisions allowing some flexibility in applying the directive in such cases. The English Mental Capacity Act 2005, for instance, states in Section 25(2)(c) (Legislation.gov.uk 2012):

An advance decision is not valid if P [i.e. a person] ... has done anything else clearly inconsistent with the advance decision remaining his fixed decision.

5.3 Revocation of Advance Directives: Conditions for Validity

Having reviewed the evidence about stability of preferences and discussed the problem of preferences changing without a concomitant change in the written provisions, we now turn to the question of how an advance directive can be validly revoked. In other words, what conditions have to be met for a revocation to have full ethical (and legal) weight? Four different aspects of advance directives are candidates for influencing the validity of a revocation—content, form, time and state of mind.

5.3.1 Content of Revocation

Logically, revoking an advance directive means that all or some of the statements it includes are negated. Depending on the content, this can imply different things. A patient with amyotrophic lateral sclerosis may write a directive refusing the future use of mechanical ventilation, only to revoke it later when becoming short of breath, thus granting his consent to mechanical ventilation. On the other hand, a patient with a schizophrenic psychosis may use an advance directive to grant anticipatory consent to acute treatment with antipsychotic medication but later change his mind (because of adverse effects experienced), revoke his decision and explicitly deny his consent to all antipsychotics leading to weight gain. These examples show that revocation can mean both giving and denying consent. Just as the validity of an advance directive does not hinge upon whether consent is given or denied, or on the form of treatment concerned, the same is true of revocation. Anything else would be inconsistent with the right to self-determination and patient autonomy.

5.3.2 Form of Revocation

Laws on advance directives often stipulate that a valid document has to be in writing, but that revocation does not have to be in a written form—for example, the law in Germany (Wiesing et al. 2010). Other laws also accept oral advance directives and do not require the written form in the first place (Office of the Maine Attorney General 2012); alternatively, the same formal conditions may be specified for both issuing and revoking a directive (Die Bundesversammlung der Schweizerischen Eidgenossenschaft 2008). The "written form" requirement can serve two aims: first, it may be introduced to heighten the level of evidence about the patient's wishes. Written statements are commonly regarded as providing stronger evidence than verbal statements reported by third parties. However, written statements can also be forged, and if a verbal statement is witnessed by more than one credible person, its evidentiary force is also robust. Moreover, there are disabled, ill or illiterate persons who are not able to write but who should still be allowed to issue and revoke an advance directive, whether by oral expression, unequivocal gestures or any other means. Whether a law requires special forms of evidence—writing, witnesses, audiovisual recording—depends on the level of mistrust in a society. In any case, it does not appear reasonable to require different standards of evidence for the advance directive and its revocation. Both are expressions of personal autonomy and, as we have seen in the previous section, both may have a pro-treatment or anti-treatment content.

The other reason why the written form is often advocated is to establish a barrier that makes people think twice before they express their preferences for future medical treatment. This barrier, then, is a form of mild paternalism. However, it has not been empirically shown that having to express something in writing really gives rise to a process of deliberation leading to a more considered and authentic expression of wishes than a verbal statement. Depending on their educational background, some individuals may have difficulty in expressing their preferences adequately in writing but are perfectly capable of doing so verbally. They may also be at a disadvantage when the information and counselling given to people who want to issue advance directives is only available in a written form. Efforts should be made to ensure that this information is as readable as possible (Mueller et al. 2010), and to also offer informational video clips, podcasts and personal counselling. If a paternalistic barrier is deemed to be justified, people should at least be offered ways of giving evidence other than the written form, such as video, dictation or personal testimony.

How can it be justified to require an advance directive to be in writing, but to allow it to be revoked orally? If patients lose their ability to write in the course of an illness but can still express themselves verbally, they could also use the other forms of quasi-written evidence mentioned above. The real reason behind this difference is probably the desire to set a higher barrier for the decision against life (which is usually the content of an advance directive) and a lower barrier for the pro-life decision (which, then, is considered to be the usual content of a revocation). It is thus normatively motivated, and this should be made transparent and openly discussed.

5.3.3 Time of Revocation

Most laws state that advance directives can be revoked "at any time" (Deutscher Bundestag 2009; Legislation.gov.uk 2012; Office of the Maine Attorney General 2012; Die Bundesversammlung der Schweizerischen Eidgenossenschaft 2008). At first sight, this seems a little strange. Why should time matter, and how is this to be understood? None of the laws specify that an advance directive can be *issued* "at any time", so why is time emphasized in the context of revocation? There may be three answers: (1) "At any time" may simply be used as a set phrase underscoring the mere fact that an advance directive can be revoked. (2) It may mark the difference from contracts, subscriptions or other legal documents that usually have a notice period (i.e. withdrawal will only take effect after a defined period). Obviously, this would not make sense: revocation is just as much an expression of wishes as issuing an advance directive; both should have immediate effect, although they cannot of course have any retroactive effect. (3) "At any time" may also refer to the various states of health or states of mind that a patient may be in when revoking an advance directive. This is discussed in the next section.

5.3.4   State of Mind

Patients may be in a different state of mind when revoking an advance directive than when they issued it; in particular, they may have lost decision-making capacity. Patients with dementia, in an acute state of delirium or stroke may still be able to utter meaningful sentences without having the cognitive preconditions for decision-making capacity. So the question arises, is decision-making capacity a necessary condition for valid revocation, just as it is for the validity of an advance directive?

This question can be answered clearly if we acknowledge that revocation is essentially nothing other than a form of advance directive itself. Revoking a directive means negating certain parts of it or the whole document and replacing it with another statement, which may be the opposite, but may also be a third option. This amounts to a new advance directive, partially or totally superseding the older one. The term "revoke" is merely a relational verb, which links the new statement to the older one, but does not have any substantive meaning on its own. Any jurisdiction will regard a more recent treatment preference as replacing an older one, even if the former does not explicitly mention and nullify the latter. As indicated at the beginning of this chapter, decision-making capacity requires an understanding of the clinical situation and the available treatment options. Whether preferences are expressed for the first time or revoked and modified, they still relate to the same clinical situation and the same treatment options. Therefore, it does not make sense to require capacity for the first expression of preferences, but not for the second one.

Yet, this might be difficult to act upon in practice. If a demented patient who had previously issued a clear, considered advance directive refusing life-sustaining measures now begs you to administer antibiotics to treat pneumonia, what should you do? If you have just withdrawn haemodialysis from a patient with end-stage renal cancer, as requested in her directive, and she then develops delusions due to uraemic encephalopathy, tears up her document and demands reinstitution of haemodialysis, what would you do? With the right communication skills, you might reach an agreement with these patients on a certain course of action, but this does not solve the ethical question, "What is the right thing to do in such a situation?" Even if we maintain that an advance directive can only be validly *revoked* by a patient with full decision-making capacity, a valid directive does not necessarily have to be implemented if it is not *applicable* to the situation. A person writing an advance directive makes certain assumptions not only about future states of health and the effects of medical treatment, but also about his or her future mental state, well-being, needs and wishes. If these assumptions turn out to have been false, an important pillar supporting the anticipatory treatment preferences crumbles, rendering the advance directive inapplicable to the current situation. Of course, this judgement of applicability is probably the most intricate challenge of dealing with anticipatory statements—all the more so if the patient cannot communicate verbally but only exhibits some kind of behaviour that may be interpreted as reflecting the patient's wishes.

5.4 Patient Behaviour Conflicting with an Advance Directive

Imagine an elderly patient suffering a severe stroke that leaves her unable to speak and paralysed on the right side of her body. In rehabilitation care, she is given food and fluids orally, as swallowing is not impaired. But every time the nurses try to give her something to eat or drink, she grumbles, presses her lips together and turns her head away. Doesn't she like the meals? Is she depressed? Is she fearful or delusional? Is this a deliberate refusal to eat and drink? Does this mean she wants to die? Should we place a feeding tube to provide artificial nutrition and hydration?

5.4.1 Practical Relevance of the Problem

These are questions of practical relevance, which are, however, rarely reflected on or systematically studied in bioethics. Two empirical studies we conducted underline this practical relevance. In an experimental study, we used hypothetical case vignettes to explore how health surrogates of dementia patients proceed when making their substitute decisions (Jox et al. 2012). We presented two vignettes about end-stage dementia patients and asked the study participants (family caregivers and professional guardians) to imagine they were the surrogates of the patients in the vignettes and were called upon to give or refuse their consent to placement of a feeding tube and insertion of a cardiac pacemaker. The vignettes contained predefined variables that are of ethical significance for surrogate decision-making (e.g. information about the patient's previous statements or current well-being). We randomly presented one of two versions of each case vignette to the participants, with the variables arguing for opposite treatment decisions. Using this method, we determined that the variable with the highest impact on surrogate treatment decisions was not the patient's previously stated preferences (as it should be according to the law), but current non-verbal behaviour (e.g. refusal of food and fluids as described above).

In an interview study involving family surrogates of patients living in a persistent vegetative state, we also found that patient behaviour matters (Kuehlmeyer et al. 2012). Even though, according to the traditional neurological view, patients in this condition do not show any form of intentional, meaningful behaviour, but only brain-stem and spinal motor reflexes and expressions of the autonomic (vegetative) nervous system, relatives interpreted these reflex and vegetative behaviours as intentional, volitional and communicative. They inferred a will to survive from the most subtle behavioural signs, and even from the fact that patients physically recovered from complications such as infections. It is not known whether such behaviour is similarly interpreted by nurses, physicians and other healthcare professionals, or whether it also affects their treatment decisions. In addition to a paucity of empirical studies, there is a lack of ethical and legal reflection on this phenomenon.

5.4.2   The "Natural Will" in German Law

German law is one of the few legal systems that recognizes such behavioural signs as an expression of a "natural will" (Jox 2006). This, however, is an extremely vague and ill-defined legal term, used in different contexts and in different ways. It probably dates back to the German philosopher Georg Wilhelm Friedrich Hegel, who used the term to refer to a deficient form of will that lacks rationality, voluntariness and consciousness (Hegel 1986). He also uses, synonymously, the terms "drive", "tendency", "passion", "physical desire", "personal interest", "opinion" and "subjective belief". In Hegel's normative system, this is a basic, selfish form of will that humans should overcome and that is clearly inferior to the free, conscious, reflective and rational will (Hegel 2001). As Hegel's philosophy significantly influenced German philosophy of law, and indeed law itself, it becomes comprehensible how this term was adopted—albeit with a different meaning—in German law. In current German (civil and criminal) law, "natural will" means any expression of will on the part of a person who does not have full decision-making capacity. Surrogates are advised not to act against the natural will; psychiatric hospitalization has to meet special conditions if it occurs against the patient's natural will. An invasive, mutilating procedure such as forced sterilization of a person lacking decisional capacity is even prohibited if there is a natural will opposing it (Jox 2006).

This concept of the "natural will" is not uncontroversial among legal scholars and ethicists. Also debated is the question whether behaviour such as that of the patient with dementia described above actually constitutes a natural will and whether this trumps an advance directive (Jox 2011). The German law on advance directives is silent on this issue, merely saying that a directive can be revoked "informally at any time" (Deutscher Bundestag 2009).

5.4.3 Tentative Classification of Patient Behaviour

In examining the ethical relevance of such patient behaviour, it is a basic desideratum to collect typical examples and develop a classification of this behavioural repertoire. I will try to address this task before going on to discuss the ethical significance of this behaviour. In an initial approach, behaviour can be categorized as *specific* or *unspecific*. Specific behaviour occurs only in certain situations and involves particular actions of an *aversive* or *appetitive* nature. Unspecific behaviour, on the other hand, is a general behavioural condition that can be observed most of the time and in many different situations.

Let us first focus on specific behaviour. As noted above, caregivers often report *food refusal* in patients with dementia (Wasson et al. 2001). In individual cases, it has been reported that patients in nursing homes push the spoon away when they are being fed, press their lips together, or turn their head away. This behaviour can be termed aversive because a specific action proposed by someone else is rejected. Another frequent case of aversive behaviour in geriatrics is refusal of medication, with patients hiding or throwing away their pills, spitting out medicines, or offering physical resistance when receiving an injection. The removal of feeding tubes by patients—a frequent occurrence—can be seen as a method of refusing both food and medication, as these tubes usually provide both. Appetitive (as opposed to aversive) behaviour, is less frequently talked about, because it causes no controversy and usually harmonizes with the care a patient is given. Examples are calling for the caregiver, grabbing food, using medication, or seeking affection and sympathy, as when patients cuddle with caregivers or hold their relatives' hands.

Let us now look at unspecific behaviour. If patients with dementia laugh a lot, sing to themselves, or are immersed in playing with various things, then this is often interpreted as a sign of pleasure, happiness and *a will to live*. On the other hand, if they do not get up in the morning, dislike eating or drinking, or ignore questions and reject communication, this can be interpreted as *a will not to live* or even a death wish. The interpretation of the will to live or die also plays an important role in the management of critically ill newborns or children. Occasionally, a surprising improvement in a small child's health is interpreted as a will to live—for instance, when a life-threatening infection is finally overcome after a few weeks. On the other hand, if against all expectations a premature infant cannot be weaned off a respirator, this might be construed by the parents and physicians as the lack of a will to live or as a death wish.

5.4.4 Careful Interpretation

The question that always arises is how specific or unspecific behaviour is to be interpreted. Let us consider the motives underlying specific behaviour. As a verbal dialogue about motives is often no longer possible, various potential interpretations have to be taken into consideration (Volicer 2004). Patients with dementia may refuse food because they are not hungry (often the case with dementia), because their sense of taste is distorted (a known side effect of medication), because chewing or swallowing is painful (possibly owing to a dry mouth or an infection), because they are suffering from undiagnosed depression (underestimated in these patients), or because they simply do not like the food that is being offered to them. Patients who remove their feeding tube may do so because it itches or hurts (a frequent consequence of inadequate skin care), because they are delusional and think that the tube is a chain (such patients often have delusional disorders), or in order to get more of the caregiver's attention. A qualitative observational study convincingly demonstrated that different caregivers interpret identical behavioural conditions of dementia patients completely differently from one another, and, as a result, deal with the patients in different ways (Pasman et al. 2003).

Let us take a step back and ask ourselves if the behavioural phenomena that have been mentioned can even be considered *expressions*—i.e. if they even serve the purpose of communicating inner mental states to the outside world. If the answer is yes, then these physical movements are more than mere physical movements in that they are regarded as transmitters of meanings and as signs referring to something outside of themselves. This link has become so integrated into our daily lives that we no longer question it, and it is hard for us to imagine that the gestures familiar to us from our daily lives could lose their standardly attached meaning as a result of illness. However, neurologists and psychiatrists know of numerous mechanisms whereby the meaning of these daily gestures is changed or entirely forfeited in cases of involuntary motor behaviour.

Let us look at motor behaviour: an end-stage dementia patient in a minimally conscious state grasping the hand of a family member does not automatically signify a desire for human contact, care and affection, but could simply be a grasp reflex, which is normal in infants and common in adults with severe damage to the frontal cortex (Alexander and Stuss 2000). If patients with dementia suck on a straw that is offered to them, this does not necessarily mean that they are thirsty, want to drink the medication or even want to stay alive; instead, it could be the effect of the sucking reflex, another sign of frontal damage. The frequent repetition of certain words can be a pathological sign, known as verbal perseveration. A patient with previous strokes, amyotrophic lateral sclerosis or antipsychotic medication may appear to laugh or cry, but in fact does not want to laugh or cry: these patterns of expression are unintentionally produced or maintained by the brainstem, independently of the person's emotional state—so-called pathological laughing and crying, or "pseudobulbar affect" (Burkey et al. 2012; Maheshwari et al. 2010). Physical movements of various sorts (e.g. epileptic phenomena, tics, dystonia or tremor) can occur without actually having to mean something. In all of these cases, the specific meaning usually borne by behavioural elements is lost as a result of a disturbance within the nervous system. It would not do the patient justice to interpret these behavioural conditions as having their normal, standardized meaning.

5.4.5 Treatment Decision Making

If in a given situation pathological, stereotypical forms of involuntary motor behaviour have been ruled out and the behaviour is indeed found to express the patient's mental state, the question remains how it should be used for decision-making about medical treatment. Can it be construed as a *performative (action-based) revocation* of an advance directive? We have seen above that revocation, like the initial writing, requires full decision-making capacity. And the behavioural expressions we introduced and classified with the aid of examples certainly do not fulfil the philosophical criteria of what we ordinarily call the human will: freedom, rationality, consciousness, ability to guide actions (Jox 2006).

On the other hand, these behavioural expressions should not be dismissed altogether. Even if they do not constitute a revocation, they may still be ethically relevant. They certainly reveal something about the patient's current well-being. They allow us to better judge whether the patient is in a happy or miserable situation, and whether certain actions or forms of treatment make them happier or more miserable. If we seek to satisfy the bioethical obligations of non-maleficence and beneficence, these behavioural signs may therefore be relevant clues (Beauchamp and Childress 2008). This assessment may be weighed against the stipulations of the advance directive. In doing so, we should be very careful to apply the patient's directive to the actual clinical and biographical situation. As has already been mentioned, advance directives involve assumptions about one's future well-being in a certain clinical situation. If these assumptions evidently prove to have been mistaken, to simply carry out the instructions would not mean to honour the patient's authentic wishes. This does not imply however, that an advance directive can be readily trumped by vague indications of supposedly false assumptions. In the end, surrogate decision-makers have to try to obtain a consistent and coherent account of the patient's wishes and well-being in order to come to a responsible decision on treatment matters.

5.5 Conclusions

Expressed treatment preferences have been found to be relatively stable, especially when they are documented in a written advance directive. The option of revocation is a crucial feature of advance directives, but revocation should be subject to the same conditions as apply to the writing of a directive. Non-verbal behaviour of patients who have lost decision-making capacity should not be construed as representing a revocation of an advance directive. It should, however, be interpreted as indicative of the patient's well-being and has to be taken into account when applying advance directives to the actual clinical situation.

Acknowledgments This chapter is partly based on earlier German articles of the author (Jox 2006, 2011). Some of the work was done as a Caroline Miles Visiting Scholar at the Ethox Centre of the University of Oxford, UK. The author thanks Dorothee Wagner von Hoff, Ph.D., for language support and Jeff Acheson for editing.

References

Alexander, M.P., and D.T. Stuss. 2000. Disorders of frontal lobe functioning. *Seminars in Neurology* 20(4): 427–437.

Appelbaum, P.S. 2007. Clinical practice. Assessment of patients' competence to consent to treatment. *The New England Journal of Medicine* 357(18): 1834–1840.

Beauchamp, T.L., and J.F. Childress. 2008. *Principles of biomedical ethics*, 6th ed. New York: Oxford University Press.

Berger, J.T., and D. Majerovitz. 1998. Stability of preferences for treatment among nursing home residents. *Gerontologist* 38(2): 217–223.

Bernat, J.L. 2001. Informed consent. *Muscle & Nerve* 24(5): 614–621.

Bosshard, G., A. Wettstein, and W. Bar. 2003. Wie stabil ist die Einstellung Betagter zu lebensverlängernden Maßnahmen? Resultate einer 3-Jahres-Katamnese von Pflegeheimbewohnern. *Zeitschrift für Gerontologie und Geriatrie* 36(2): 124–129.

Burkey, M.D., C. Howell, M.A. Riddle, and B.J. Coffey. 2012. Antipsychotic-induced psychotic-like syndrome and pathological smiling in an adolescent. *Journal of Child and Adolescent Psychopharmacology* 22(1): 92–95.

Carmel, S., and E.J. Mutran. 1999. Stability of elderly persons' expressed preferences regarding the use of life-sustaining treatments. *Social Science & Medicine* 49(3): 303–311.

Danis, M., J. Garrett, R. Harris, and D.L. Patrick. 1994. Stability of choices about life-sustaining treatments. *Annals of Internal Medicine* 120(7): 567–573.

Davis, J.K. 2002. The concept of precedent autonomy. *Bioethics* 16(2): 114–133.

Deutscher Bundestag. 2009. Drittes Gesetz zur Änderung des Betreuungsrechts. *Bundesgesetzblatt* 48: 2286–2287.

Die Bundesversammlung der Schweizerischen Eidgenossenschaft. 2008. *Schweizerisches Zivilgesetzbuch (Erwachsenenschutz, Personenrecht und Kindesrecht) Änderung vom 19. Dezember 2008.* Available from http://www.admin.ch/ch/d/ff/2009/141.pdf.

Ditto, P.H., W.D. Smucker, J.H. Danks, J.A. Jacobson, R.M. Houts, A. Fagerlin, K.M. Coppola, and R.M. Gready. 2003. Stability of older adults' preferences for life-sustaining medical treatment. *Health Psychology* 22(6): 605–615.

Ditto, P.H., J.A. Jacobson, W.D. Smucker, J.H. Danks, and A. Fagerlin. 2006. Context changes choices: A prospective study of the effects of hospitalization on life-sustaining treatment preferences. *Medical Decision Making* 26(4): 313–322.

Edwards, S.J. 2005. Research participation and the right to withdraw. *Bioethics* 19(2): 112–130.

Emanuel, L.L. 2008. Advance directives. *Annual Review of Medicine* 59: 187–198.

Emanuel, L.L., E.J. Emanuel, J.D. Stoeckle, L.R. Hummel, and M.J. Barry. 1994. Advance directives. Stability of patients' treatment choices. *Archives of Internal Medicine* 154(2): 209–217.

Emanuel, E.J., D. Wendler, and C. Grady. 2000. What makes clinical research ethical? *Journal of the American Medical Association* 283(20): 2701–2711.

Everhart, M.A., and R.A. Pearlman. 1990. Stability of patient preferences regarding life-sustaining treatments. *Chest* 97(1): 159–164.

Faden, R.R., T.L. Beauchamp, and N.M.P. King. 1986. *A history and theory of informed consent.* New York: Oxford University Press.

Froman, R.D., and S.V. Owen. 2005. Randomized study of stability and change in patients' advance directives. *Research in Nursing & Health* 28(5): 398–407.

Hegel, G.W.F. 1986. *Werke in 20 Bänden und Register, Bd. 3: Philosophie des Geistes.* 8. Aufl. ed. Frankfurt am Main: Suhrkamp.

Hegel, G.W.F. 2001. *Werke in 20 Bänden und Register, Bd. 17: Vorlesungen über die Philosophie der Religion.* 5. Aufl. ed. Frankfurt am Main: Suhrkamp.

Henderson, C., J.W. Swanson, G. Szmukler, G. Thornicroft, and M. Zinkler. 2008. A typology of advance statements in mental health care. *Psychiatric Services* 59(1): 63–71.

Inthorn, J. 2008. Wünsche und Befürchtungen von Patienten bei der Einrichtung von Patientenverfügungen. Ergebnisse einer Studie zum Patientenverfügungsgesetz in Österreich. *Bulletin de la Société des Sciences Médicales du Grand-Duché de Luxembourg* 3: 429–440.

Johnston, C., and J. Liddle. 2007. The Mental Capacity Act 2005: A new framework for healthcare decision making. *Journal of Medical Ethics* 33(2): 94–97.

Jox, R.J. 2006. Der "natürliche Wille" als Entscheidungskriterium: Rechtliche, handlungstheoretische und ethische Aspekte. In *Entscheidungen am Lebensende in der modernen Medizin: Ethik, Recht, Ökonomie und Klinik*, ed. J. Schildmann, U. Fahr, and J. Vollmann. Berlin: LIT Verlag.

Jox, R.J. 2011. Widerruf der Patientenverfügung und Umgang mit dem natürlichen Willen. In *Patientenverfügung. Das neue Gesetz in der Praxis*, ed. G.D. Borasio, H.-J. Heßler, R.J. Jox, and C. Meier. Stuttgart: Kohlhammer.

Jox, R.J., M. Krebs, J. Bickhardt, K. Hessdorfer, S. Roller, and G.D. Borasio. 2008. How strictly should advance decisions be followed? The patients' opinion. *Palliative Medicine* 22(5): 675–676.

Jox, R.J., E. Denke, J. Hamann, R. Mendel, H. Forstl, and G.D. Borasio. 2012. Surrogate decision making for patients with end-stage dementia. *International Journal of Geriatric Psychiatry* 27 (10): 1045–1052. Epub 2011 Dec 5.

Koertner, U., C. Kopetzki, and M. Kletecka-Pulker. 2007. *Das österreichische Patientenverfügungsgesetz 2006: Ethische und rechtliche Aspekte*, vol. Schriftenreihe Ethik und Recht in der Medizin. Vienna/New York: Springer.

Krumholz, H.M., R.S. Phillips, M.B. Hamel, J.M. Teno, P. Bellamy, S.K. Broste, R.M. Califf, H. Vidaillet, R.B. Davis, L.H. Muhlbaier, A.F. Connors Jr., J. Lynn, and L. Goldman. 1998. Resuscitation preferences among patients with severe congestive heart failure: Results from the SUPPORT project. Study to Understand Prognoses and Preferences for Outcomes and Risks of Treatments. *Circulation* 98(7): 648–655.

Kuehlmeyer, K., G.D. Borasio, and R.J. Jox. 2012. How family caregivers' medical and moral assumptions influence decision making for patients in the vegetative state: A qualitative interview study. *Journal of Medical Ethics* 38(6): 332–337.

Kutner, L. 1969. Due process of euthanasia: The living will, a proposal. *Indiana Law Journal* 44: 549.

Landis, J.R., and G.G. Koch. 1977. The measurement of observer agreement for categorical data. *Biometrics* 33(1): 159–174.

Legislation.gov.uk. *Mental Capacity Act 2005*. The National Archives 2012. Available from http://www.legislation.gov.uk/ukpga/2005/9/contents

Loewenstein, G. 2005. Hot-cold empathy gaps and medical decision making. *Health Psychology* 24(4 Suppl): S49–S56.

Maheshwari, S., A. Figueiredo, and A. Goel. 2010. Pathological crying as a manifestation of spontaneous haemorrhage in a pontine cavernous haemangioma. *Journal of Clinical Neuroscience* 17(5): 662–663.

Martin, V.C., and K.A. Roberto. 2006. Assessing the stability of values and health care preferences of older adults: A long-term comparison. *Journal of Gerontological Nursing* 32(11): 23–31; quiz 32–3.

McConnell, T. 2010. The inalienable right to withdraw from research. *The Journal of Law, Medicine & Ethics* 38(4): 840–846.

McParland, E., A. Likourezos, E. Chichin, T. Castor, and Be.Be. Paris. 2003. Stability of preferences regarding life-sustaining treatment: A two-year prospective study of nursing home residents. *The Mount Sinai Journal of Medicine* 70(2): 85–92.

Mueller, L.A., K.I. Reid, and P.S. Mueller. 2010. Readability of state-sponsored advance directive forms in the United States: A cross sectional study. *BMC Medical Ethics* 11: 6.

Muthappan, P., H. Forster, and D. Wendler. 2005. Research advance directives: Protection or obstacle? *The American Journal of Psychiatry* 162(12): 2389–2391.

Nicholas, L.H., K.M. Langa, T.J. Iwashyna, and D.R. Weir. 2011. Regional variation in the association between advance directives and end-of-life Medicare expenditures. *Journal of the American Medical Association* 306(13): 1447–1453.

Office of the Maine Attorney General. 2012. *Advance directive/Living will*. Available from http://www.maine.gov/ag/elder_issues/living_will.shtml

Pasman, H.R., B.A. The, B.D. Onwuteaka-Philipsen, G. van der Wal, and M.W. Ribbe. 2003. Feeding nursing home patients with severe dementia: A qualitative study. *Journal of Advanced Nursing* 42(3): 304–311.

Pruchno, R.A., M.J. Rovine, F. Cartwright, and M. Wilson-Genderson. 2008. Stability and change in patient preferences and spouse substituted judgments regarding dialysis continuation. *The Journals of Gerontology. Series B, Psychological Sciences and Social Sciences* 63(2): S81–S91.

Rich, B.A. 1997. Prospective autonomy and critical interests: A narrative defense of the moral authority of advance directives. *Cambridge Quarterly of Healthcare Ethics* 6(2): 138–147.

Rich, B.A. 2004. Current legal status of advance directives in the United States. *Wiener Klinische Wochenschrift* 116(13): 420–426.

Schaden, E., P. Herczeg, S. Hacker, A. Schopper, and C.G. Krenn. 2010. The role of advance directives in end-of-life decisions in Austria: Survey of intensive care physicians. *BMC Medical Ethics* 11: 19.

Schaefer, G.O., and A. Wertheimer. 2010. The right to withdraw from research. *Kennedy Institute of Ethics Journal* 20(4): 329–352.

Schwartz, C.E., M.P. Merriman, G.W. Reed, and B.J. Hammes. 2004. Measuring patient treatment preferences in end-of-life care research: Applications for advance care planning interventions and response shift research. *Journal of Palliative Medicine* 7(2): 233–245.

Silveira, M.J., S.Y. Kim, and K.M. Langa. 2010. Advance directives and outcomes of surrogate decision making before death. *The New England Journal of Medicine* 362(13): 1211–1218.

Srebnik, D.S., L.T. Rutherford, T. Peto, J. Russo, E. Zick, C. Jaffe, and P. Holtzheimer. 2005. The content and clinical utility of psychiatric advance directives. *Psychiatric Services* 56(5): 592–598.

Stocking, C.B., G.W. Hougham, D.D. Danner, M.B. Patterson, P.J. Whitehouse, and G.A. Sachs. 2006. Speaking of research advance directives: Planning for future research participation. *Neurology* 66(9): 1361–1366.

Volicer, L. 2004. Dementias. In *Palliative care in neurology*, ed. R. Voltz, J.L. Bernat, and G.D. Borasio. Oxford: Oxford University Press.

Wasson, K., H. Tate, and C. Hayes. 2001. Food refusal and dysphagia in older people with dementia: Ethical and practical issues. *International Journal of Palliative Nursing* 7(10): 465–471.

Weissman, J.S., J.S. Haas, F.J. Fowler Jr., C. Gatsonis, M.P. Massagli, G.R. Seage 3rd, and
    P. Cleary. 1999. The stability of preferences for life-sustaining care among persons with
    AIDS in the Boston Health Study. *Medical Decision Making* 19(1): 16–26.
Wiesing, U., R.J. Jox, H.-J. Heßler, and G.D. Borasio. 2010. A new law on advance directives in
    Germany. *Journal of Medical Ethics* 36: 779–783.
Wittink, M.N., K.H. Morales, L.A. Meoni, D.E. Ford, N.Y. Wang, M.J. Klag, and J.J. Gallo. 2008.
    Stability of preferences for end-of-life treatment after 3 years of follow-up: The Johns Hopkins
    Precursors Study. *Archives of Internal Medicine* 168(19): 2125–2130.
World Medical Association. 2008. *Declaration of Helsinki – Ethical principles for medical
    research involving human subjects*. Available from http://www.wma.net/en/30publications/
    10policies/b3/

Chapter 6
Limitations to the Scope and Binding Force of Advance Directives: The Conflict Between Compulsory Treatment and the Right to Self-Determination

Jacqueline M. Atkinson and Jacquie Reilly

6.1 Introduction

Advance planning for people with mental illness involves allowing competent individuals to specify their treatment preferences in advance of future incapacity. Supporters of advance planning argue that it can enhance autonomy and allow people to participate in their future treatment decisions, allowing patients and clinicians to engage in a constructive approach to treatment planning, while opponents claim that advance directives are problematic both legally and ethically and not workable in practice (Morrissey 2010).

In all areas of medicine, consideration will have to be given to how local capacity (or competence) legislation influences the determination of capacity to make an advance directive. In psychiatry, however, advance directives are also "nested in larger structures of mental health law and policy that protect the interests of parties other than the patient" (Swanson et al. 2006). In only a few jurisdictions are advance directives (or their equivalent) written into mental health law itself. The issue here is whether mental health law can override a competently made advance directive, and if so, in what circumstances.

The different types of advance directive will also influence how useful they are based on different dimensions. They can be strong or weak (or legally binding or not), opt-in or opt-out directives, or independent or cooperative (agreed with staff) (Atkinson and Garner 2003). What is allowed in the directive will also affect use and will be limited by legislation or policy. Thus, the directive may only cover treatment for a mental illness or allow for medical treatment of other conditions when the person is mentally unwell and unable to make a decision. In many

J.M. Atkinson, Ph.D., CPsychol (✉)
Public Health, University of Glasgow, Glasgow, UK
e-mail: jacqueline.reilly@glasgow.ac.uk

J. Reilly
Public Health, University of Glasgow School of Medicine, Glasgow, UK

P. Lack et al. (eds.), *Advance Directives*, International Library of Ethics, Law, and the New Medicine 54, DOI 10.1007/978-94-007-7377-6_6,
© Springer Science+Business Media Dordrecht 2014

jurisdictions, this would require two different advance directives. Even in respect of treatment for a mental illness, the definition of what "treatment" (even "medical treatment") may cover will vary. This allows for inpatient nursing care to be defined as "treatment" to justify someone needing to be detained in hospital to receive "treatment".

There have been suggestions that some people would like to see advance directives incorporate much wider concerns, such as child care, pet care, financial management (ranging from paying bills to confiscation of credit cards when manic) and housing issues (including maintaining tenancy) (Atkinson et al. 2004). This idea is generally not popular with clinical staff, who do not see these issues as part of their remit (although some might be appropriate for other staff, such as social workers in the case of either child care or financial/housing issues). In addition, very clear legislative authority would be required, for example, to refuse to allow someone access to their credit card.

Managing these wider issues might be easier, and possibly more appropriate, where the advance directive allows for the appointment of a proxy or similar. Again, legal jurisdictions vary considerably on this point.

In Scotland, an attempt was made in the Mental Health (Care and Treatment) (Scotland) Act 2003 to manage these two aspects of advance planning. *Advance statements* were introduced to cover decisions about treatment for a mental illness, and *personal statements*—which do not have the same legal standing—to cover the wider range of a person's wishes (Scottish Executive 2004). Although this may make sense in theory, things have not been so clear in practice. In part, this is because what is covered by the term "treatment" is not clearly defined. Thus, if someone says they want to jog for an hour a day as this helps keep them calm and reduces their agitation, thereby also reducing their need for medication, does this count as treatment? Where staff or other resources are involved, the issue may become complex. One consequence of this division is that some voluntary organizations whose staff help patients to write independent advance statements are incorporating these wishes in the advance statement, on the grounds that the mental health tribunal, when making decisions under the Mental Health (Care and Treatment) (Scotland) Act, has to consider whatever is in an advance statement. This may then mean that more advance statements, or parts of statements, are overturned by the tribunal, which in turn leads people to believe that such statements are not being considered seriously, or followed, and that therefore there is no point in making one.

Despite a lot of early interest in advance directives, including from user groups, the uptake has been comparatively low. It should be remembered that they are only really useful for those who are going to be subject to mental health legislation—i.e. those assessed as fitting the criteria for detention or compulsory treatment (e.g. a lack of capacity, or risk to self or others)—who are likely to be suffering from illnesses such as schizophrenia, bipolar disorder or major depression. Therefore, they are not relevant to the majority of psychiatric patients, who are treated voluntarily and do not come under any law. Given that it is unlikely that any jurisdiction would allow a potentially detainable patient to refuse to be detained if this means that that they are free to harm others (or, in some jurisdictions, themselves, e.g. through self-neglect), an acceptable means of advance planning may have to be found. The fluctuating nature of mental illness in itself contributes to the difficulties involved in writing and implementing advance directives, but they remain one way for the person to make their wishes known. Whatever the case, there will always be a conflict between self-determination, approaches to capacity and society's wishes to protect both itself and those deemed vulnerable.

6.2 Compulsory Treatment

Compulsory treatment is a complicated and sometimes controversial issue in psychiatry, which has been debated for a number of decades (Agnetti 2008). There are three lines of argument in defence of coercive treatment: (1) societal interests in protecting others, (2) the patients' own health interests and (3) patient autonomy. Sjöstrand and Helgesson (2008) argue that:

> Coercive treatment may be required in order to promote the patient's health interests, but health interests have to waive if they go against the autonomous interests of the patient. [...] non-autonomous patients can have reasons, rooted in their deeply-set values, to renounce compulsory institutional treatment, and [...] such reasons should be respected unless it can be assumed that their new predicaments have caused them to change their views.

Forced treatment in this sense is seen as an erosion of autonomy. The question then is, if a person with a mental disorder claims not to want medical treatment, who should decide whether they should be treated against their will when their decision was made when rational? To what extent should an advance directive have force in a situation where, for example, an individual has dementia and (arguably) is becoming a different person from the one who instigated the directive in the first place?

It has been argued that notions of self-determination assume the presence of a persistent unitary being, but with some psychiatric disorders there can be a marked qualitative change to the person, raising questions about which person should be able to make decisions. While the well person has the right to self-determination, notions of the changing "self" need to be taken into consideration in this context.

The potentially changing nature of self is important here. A patient with cancer who refuses treatment is unlikely to be challenged that the disease has somehow caused them to become a different person; however, the cancer may have caused them to re-evaluate some previously held values and attitudes. Thus, it might be accepted that dying is better than the side effects of treatment. For a person with a mental illness, it is less likely to be accepted that being acutely ill, or even dying, is better than the side effects of treatment. There may be two reasons for this. One is the argument that the mental illness has affected the person's capacity to make decisions. The other is the question of resource use. In physical illness, most refusal of treatment is in relation to patients with a terminal illness, and there will be an end to the call on hospital and other resources. This is unlikely to be the case for people who refuse treatment for mental illness, where ongoing care might be both difficult and protracted (see the discussion in Sect. 6.4). Even in jurisdictions which allow a person to opt out of treatment, nowhere is a person allowed to opt out of being detained in hospital. In physical health, although a person may be able to refuse all treatment, in some circumstances (particular infectious diseases) they may be able to be held in hospital to protect the public. Similarly, although some jurisdictions allow for an opt-in to treatment, or specification of a preferred treatment, nowhere does the law allow a patient to demand a treatment against the advice of a clinician.

6.3 Capacity and Advance Directives in Psychiatry

An advance statement is made at a time when a person has capacity (is competent), to come into effect when they are incapacitated. In many cases, this loss of capacity will be permanent. In psychiatry, it is more likely to be fluctuating, and it may only affect part of a person's decision-making ability. This raises questions about when it is made, when it is invoked and when it can be changed. In Scottish law, a form of impaired capacity known as "significantly impaired decision-making ability" was introduced, relating only to the person's ability to make medical decisions about treatment for their mental illness (Scottish Parliament 2003).

How and when the decision to invoke an advance directive is made, and by whom, can become subject to debate. Again, variations in the law will have an impact. Everywhere, not everyone who becomes incompetent will be subject to detention or compulsory treatment. In some places, the law requires some form of incapacity before a person can be compulsorily treated, whilst elsewhere it is possible for a competent person to undergo compulsory treatment. Advance directives may be used at such times to persuade a person to follow a previously planned course of action, but they would seem to raise significant legal issues if enforced in the face of a competent person making a different choice. It would seem that some "formal" determination of capacity/competence is needed, even if this does not involve a court and legal decision.

There are different opinions on whether an advance directive can be revoked at any time or only when a person is competent. Whilst it would seem reasonable that the decision to revoke an advance directive should only be made by the person when capable—otherwise, what is the point of the directive?—the case is not so clear-cut (Atkinson 2007). In the scenario where only a capable person can revoke an advance directive, there is an assumption that the incapable person is somehow "less" and has less right to have their wishes adhered to. "Less" here is taken to mean that the person has lost something of themselves, not that they are a lesser human being.

If a person makes an advance directive agreeing to treatment when they are ill because they know, from past experience, that they will refuse treatment when ill, allowing them to then revoke that advance directive when ill negates the point of it. If, however, a person has refused treatment, and this is honoured, but the illness lasts much longer than in previous cases, and the person—albeit reluctantly—decides that taking medication is better than remaining in hospital/remaining ill, then, by the same argument, they would not be able to revoke their advance directive and could find themselves locked into a no-treatment scenario. Whether a philosophical case can be made for a distinction between these two cases is debatable, but a practical one certainly can be. In both cases, the desire is to preserve or regain the "well" person. Just as a person with a "do not resuscitate" order who indicates that they want to live would have this wish supported, it could be argued that a wish to regain health can be supported in the same way. Thus, life and health will always "trump" death and illness.

Where compulsory detention or treatment is needed, questions arise as to how to balance a patient's right to autonomy and self-determination with a physician's duty of care. There is undoubtedly an inherent conflict and a balance to be struck between the two. How do you balance the rights of an individual with the clinical need for treatment in psychiatry? The main issues that arise relate to the capacity of an individual to make an advance directive, when a directive is written and what it actually contains.

6.4   Outcomes of Advance Directives in Psychiatry

As in other specialties, advance directives support or enhance autonomy in the sense of encouraging the independence of the individual, both in making realistic choices around preferred treatment or consequences of reduced treatment and in accepting the responsibility for these decisions within a service framework which increasingly looks for the participation of patients in their treatment and management (Halpern and Szmukler 1997; Srebnik and Brodoff 2003). Enhancing communication between patient and physician (or other health professional) benefits both parties, as it encourages them to fully explain the reasons for their decisions, which may contribute to more appropriate or better management and treatment in the future.

Since a psychiatric advance directive will only come into force when a person has lost capacity, the population in question comprises those with a mental illness in which capacity may become impaired, rather than the wider category of mental health problems. This significantly reduces the number of people who may want to make advance directives and helps to clarify their purpose and remit.

To argue that advance directives are useful in psychiatry means being clear about what they are intended to achieve and thus what outcome measures might be appropriate. Outcome measures may best be suggested by both the type of advance directive involved and the philosophy behind it. Autonomy and empowerment, for example, would be best assessed by asking patients whether having the plan had made them feel more involved or in control when detained or admitted. The drawing-up of joint plans could include assessment of whether each party has a better understanding of the other and whether trust or other aspects of the physician–patient relationship had been improved. According to a survey of various stakeholders (Backlar et al. 2001), other possible benefits of advance directives include promoting responsibility in patients and having an impact on stigma. Research which only focuses on quantitative measures, such as length of hospital stay or use of legal procedures, may miss the point of someone who wishes to make their wishes known, even if they are not then acted on.

The resource issue may not be important to the individual, but it may be important to health services, and indeed the wider society which may have to fund long-term care. This situation has occurred in the Canadian province of Ontario, where in one case a patient had avoided regular treatment with antipsy-chotic medication for over 25 years by initiating multiple legal challenges and, during that time, had been detained in various psychiatric facilities, with all the resultant resource implications (O'Reilly et al. 2009). Other cases in the area had different outcomes.

O'Reilly et al. (2007) identified six cases of judicial support for refusal of psychiatric treatment during a 15-year study period. The courts in Ontario overturned a review board's confirmation of a physician's finding of treatment incapacity in only two cases. In four cases, the court held that the person could not be treated because, while capable, they had previously expressed a wish not to be treated. Only one patient was discharged from hospital without treatment. Despite earlier court decisions upholding their right to refuse treatment, four patients were treated over their objections. These patients were detained untreated in hospital for lengthy periods of time (between 5 and 10 years) and only became well enough to be discharged when treatment was given. Other jurisdictions in Canada do not use such legally binding contracts, and as regards the situation in Ontario the authors comment:

> We would presume that prolonged and unnecessary deprivation of a patient's liberty would be viewed as not in the patient's best interest. The approach used in Manitoba and Nova Scotia views an expressed wish as nonbinding if it endangers the health or safety of the patient or others. This presumably could be used to ensure that a committed patient, who is at significant risk, receives treatment that may lead to the patient's release.

The question here is whether patients might not have been better served if their preferred choices had been overridden by the courts, given their lengthy detention.

6.5 Advance Directives in Practice in Psychiatry

Many practical issues are involved in the use of advance directives in mental health care, most of them transcending jurisdiction (Srebnik and Brodoff 2003). They include: where the directives will be stored and accessed; how and when directives are activated; competence to make and revoke directives; and the content of directives, including where they may not be compatible with standards of care or raise legal concerns regarding liability (for honouring or not honouring directives).

There appear to be widespread concerns regarding the impacts of directives on resources and resource allocation.

One resource-related issue is the time required by members of staff involved in drawing up advance directives where this is either necessary or preferred by patients. Although the issue will not arise in Scotland in the same way, the question has been asked in the US whether such time was "billable"? (Srebnik and Brodoff 2003). People are right to be concerned about the availability and accessibility of directives. This—and staff's knowledge of them—seems to be a problem, even when the directive is included in the case notes (Papageorgiou et al. 2002, 2004). Interest in advance directives on the part of staff may be a vital component in their use and success.

The content of advance directives also raises questions with regard to imple-mentation. Although in some jurisdictions they clearly relate to clinical or healthcare decisions, these can be widely as well as narrowly interpreted. There is also evidence that some people would like to see advance directives incorporate matters other than medical decisions, including financial and housing issues (Atkinson et al. 2004). Managing these wider issues might be easier, or more appropriate, where the advance directive allows for the appointment of a proxy or similar; however, such issues may be less manageable where there is no clear person with the necessary responsibility or legal powers. Even where a proxy is appointed, there is still the question of whether they act on "substituted judgement"—i.e. doing what (they think) the patient would do in the circumstances—or in the patient's best interest. This often means acting in what others consider to be the patient's best interest, which may be against the patient's wishes. Thus the proxy may decide, in the face of treatment refusal, that it is still in the patient's best interest for them to be treated. If self-determination is totally supported, then it could be argued that doing what the person wants is, by definition, in their best interest.

The concern that large numbers of patients will opt out of treatment does not appear to have been borne out in places where advance directives have been legally sanctioned for some time. For example, in a study involving 40 patients in the US state of Oregon—where the appointment of a surrogate decision-maker is permitted in a psychiatric advance directive—no directives refused all treatment, although 57 % refused electroconvulsive therapy and 27 % refused the antipsychotic halo-peridol (Backlar et al. 2001). In the US, where there are different thresholds for commitment in various states, there appears to have been at least as much interest in opting in to treatment as opting out. A thematic analysis of 55 advance statements which came before a tribunal hearing in Scotland showed that, although 96 % of statements included at least one specific treatment refusal, 45 % also named specific medications the patients were prepared to take. The main treatment refused was electroconvulsive therapy (42 %), and 25 % did not want to receive depot injections. In some cases, reasons—usually relating to adverse effects—were given for refusal of a particular medication. Other refusals concerned group therapy (three people) and neurosurgery (two people). Where people named specific drugs they were prepared to take, it was usually mentioned that these had worked well in the past. A number of people requested other activities, usually with a reference to preventing boredom (Reilly and Atkinson 2010).

A further issue is the numbers of people who actually make a directive. Srebnik et al. (2003) asked heavy users of crisis services and hospitalization if they were interested in making an advance directive, with just over half agreeing. The main reasons given were that their case manager suggested it or a general belief that it would prove useful. Only 8 % of those who gave a reason explicitly stated that they wanted to plan for a time when they lacked capacity. Other patient characteristics found by these researchers to be positively related to having an interest in making an advance directive were as follows: case manager support, major depression, not having schizophrenia, and no hospital or outpatient commitment orders in the past 2 years. The main reasons given for not wanting to make an advance directive were that the person did not believe they had a mental illness or that they were likely to become ill again, with about the same number being satisfied with the plans currently in place to manage future illness or crises. A similar finding was reported in another US study (Backlar et al. 2001), where 20 % wanted their physician's treatment decisions to stand and 20 % wanted current treatment to be continued. This is an important point if uptake of advance plans is taken as one measure of success. Not everyone may see the need for them, but this may mean that services are meeting patients needs or that good relationships exist between patients and their physicians, rather than the reverse.

Not all people who do not want to make advance directives, however, are satisfied with services, and there is evidence that some people see no point, as they believe they will be ignored. In the study by Backlar et al. (2001), although 88 % of service users were satisfied with their advance directive at 8–10 months' follow-up, almost half expressed reservations, and they were generally less positive and more critical than they had been. The patients' surrogates and service providers reported that the advance directive had had little impact. La Fond and Srebnik (2002) discuss the potential impact that advance directives may have on patients' perceptions of coercion in a US setting, but they offer no evidence. At 1-year follow-up in a London study (Papageorgiou et al. 2004) using the *Preferences for Care* booklet (74 % of patients contacted), only 9 of 59 (15 %) said it had been helpful. Nevertheless, 41 % would want to use it again and 44 % said they would recommend it to others. A view often encountered is that one of the reasons for the low interest shown in making advance directives is that patients do not believe their wishes will be adhered to, so they see no point. Taken together, these findings would suggest that the views of professionals on advance directives might be particularly important if there is to be appropriate uptake.

In a UK postal survey (Atkinson et al. 2004), over 450 stakeholders were asked for their views on five models of advance directives. The majority of all groups except psychiatrists responded positively to the question "Do you think we need advance directives?" (voluntary organizations 89 %, Directors of Social Work/ Social Services 82 %, community psychiatric nurses 79 %, Directors of Mental Health Trusts 71 %, mental health officers 66 %, and psychiatrists 28 %).

Psychiatrists were significantly less likely than other groups to want to work with opt-out models that would allow patients to refuse treatment.

6.6 Conclusions

Few legal jurisdictions allow a patient to refuse all treatment for a mental illness, without the option for this to be overturned by clinicians, and nowhere can a person refuse to be detained in hospital, even if they are then not treated. The person's right to self-determination is complicated in psychiatry, where questions arise as to the ill person's capacity to make decisions which override those of the capacitous (well) person—for example, the well person wanting and the ill person refusing treatment. Where the well person wants to refuse treatment, questions arise about the ongoing care of that person when they become acutely ill, especially where this may last for a prolonged period.

Given that it is unlikely that any jurisdiction would allow a potentially detainable patient to refuse to be detained if this means that that they are free to harm others (or in some jurisdictions, themselves, e.g. through self neglect), an acceptable means of advance planning may have to be found. The fluctuating nature of mental illness in itself contributes to the difficulties involved in writing and implementing advance directives, but they remain one way for the person to make their wishes known. Whatever the case, there will always be a conflict between self determination, approaches to capacity and society's wishes to protect both itself and those deemed vulnerable.

References

Agnetti, G. 2008. The consumer movement and compulsory treatment: A professional outlook. *International Journal of Mental Health* 37(4): 33–45.

Atkinson, J.M. 2007. *Advance directives in psychiatry: Theory, practice and ethics*. London: Jessica Kingsley.

Atkinson, J.M., and H.C. Garner. 2003. Advance directives in mental health. *Psychiatric Bulletin* 27: 437.

Atkinson, J.M., H.C. Garner, and W. Harper Gilmour. 2004. Models of advance directives in mental health care: Stakeholder views. *Social Psychiatry and Psychiatric Epidemiology* 39(8): 673–680.

Backlar, P., B.H. McFarland, J.W. Swanson, and J. Mahler. 2001. Consumer, provider, and informal caregiver opinions on psychiatric advance directives. *Administration and Policy in Mental Health* 28: 427–441.

Halpern, A., and G. Szmukler. 1997. Psychiatric advance directives: Reconciling autonomy and non-consensual treatment. *The Psychiatrist* 21: 323–327.

La Fond, J., and D. Srebnik. 2002. The impact of mental health advance directives on patient perceptions of coercion in civil commitment and treatment decisions. *International Journal of Law and Psychiatry* 25(6): 537–555.

Morrissey, F. 2010. Advance directives in mental health care: Hearing the voice of the mentally ill. *Medico Legal Journal of Ireland* 16(1): 21.

O'Reilly, R., J. Grey, and S. State. 2009. An exchange of views on what constitutes reasonable review of treatment capacity. *Research Insights* 6(2): 1–14.

Papageorgiou, A., M. King, A. Janmohamed, O. Davidson, and J. Dawson. 2002. Advance directives for patients compulsorily admitted to hospital with serious mental illness: Randomised controlled trial. *The British Journal of Psychiatry* 181: 513–519.

Papageorgiou, A., A. Janmohamed, M. King, O. Davidson, and J. Dawson. 2004. Advance directives for patients compulsorily admitted to hospital with serious mental disorders: Directive content and feedback from patients and professionals. *Journal of Mental Health* 13: 379–388.

Reilly, J., and J. Atkinson. 2010. The content of mental health advance directives: Advance statements in Scotland. *International Journal of Law and Psychiatry* 33(2): 116–121.

Scottish Executive. 2004. *The new mental health act: A guide to advance statements*. Edinburgh: Scottish Executive.

Scottish Parliament. 2003. *Mental health (care and treatment) (Scotland) act*. Edinburgh: Stationery Office.

Sjöstrand, M., and G. Helgesson. 2008. Coercive treatment and autonomy in psychiatry. *Bioethics* 22(2): 113–120.

Srebnik, D., and L. Brodoff. 2003. Implementing psychiatric advance directives: Service provider issues and answers. *The Journal of Behavioural Health Services & Research* 30: 253–268.

Srebnik, D.S., J. Russo, J. Sage, T. Peto, and E. Zick. 2003. Interest in psychiatric advance statements among high users of crisis services and hospitalistion, *paychiatric services* 54: 330–336.

Swanson, J.W., M.S. Swartz, E.B. Elbogen, R.A. Van Dorn, J. Ferron, H.R. Wagner, B.J. McCauley, and M. Kim. 2006. Facilitated psychiatric advance directives: A randomized trial of an intervention to foster advance treatment planning among persons with severe mental illness. *The American Journal of Psychiatry* 34: 43–57.

Chapter 7
Advance Directives in the Context of Imprisonment

Bernice S. Elger

7.1 Introduction

In most jurisdictions, patients have successfully fought for their right to self-determination (Cox and Sachs 1994). In many countries, healthcare personnel must respect treatment refusals of competent patients. Advance directives are legally recognized in many countries as a measure allowing patients to anticipate future situations of incompetence and to extend self-determination to these situations, either by preparing a "living will" concerning future treatments or by designating a surrogate to make decisions on their behalf (Goffin 2012; Emanuel 2008).

Healthcare in the context of imprisonment is guided by the principle of equivalence. Enshrined in international soft law, recommendations from the European Committee for the Prevention of Torture and decisions of the European Court of Human Rights (Council of Europe 1998; CPT 2006; United Nations 1982, 1990; Elger 2008a; European Court of Human Rights 1994), this principle requires that detainees have access to the same standard of healthcare that is available outside prisons[1] in the country concerned. The principle of equivalence extends to ethical requirements such as informed consent, confidentiality and, more generally, the right to self-determination in relation to health interventions.

In line with the principle of equivalence, it therefore seems straightforward to grant patients in the context of detention the same autonomy rights regarding future treatments as apply outside prisons. In what follows, we will focus on the legal

---

[1] In this chapter, the term "prison" (as well as the term "prisoner") is used in the British sense and refers to any kind of detention facility (or person detained in such a facility), including jails (the US term for remand prisons).

B.S. Elger (✉)
Institute for Biomedical Ethics, University of Basel, Basel, Switzerland

Centre for Legal Medicine, University of Geneva, Geneva, Switzerland
e-mail: b.elger@unibas.ch; bernice.elger@unige.ch

P. Lack et al. (eds.), *Advance Directives*, International Library of Ethics, Law, and the New Medicine 54, DOI 10.1007/978-94-007-7377-6_7,
© Springer Science+Business Media Dordrecht 2014

framework existing in the Swiss canton of Geneva, where the principle of equivalence was confirmed in a cantonal decree in 2000 (Elger 2008c, 2011) and respect for advance directives has been part of cantonal health laws for many years (Canton de Genève 2006). The Swiss Civil Code has recently been revised, and the new adult protection law which came into force at the beginning of 2013 makes advance directives binding in all cantons (Swiss Confederation 2012).

In the prison context, advance directives typically play a significant role in three situations:

1. Hunger strikers might wish to indicate whether they agree to resuscitation and feeding once they have become incompetent (Guilbert 2001; Sebo et al. 2004).
2. Dying prisoners may wish to ensure that end-of-life care is in line with their preferences when they lose legal capacity to make decisions on their own (Linder and Meyers 2007; Dubler 1998).
3. Prisoners with mental disorders might wish to exercise self-determination and increase their own control over the treatments they receive in psychiatric emergencies by executing advance directives. This may involve authorizing future treatments even over their subsequent resistance (Monahan et al. 2003).

This chapter will first examine general ethical and legal aspects of advance directives in the context of imprisonment and then discuss advance directives in the three above-mentioned situations—hunger strikes, end-of-life care and psychiatric treatment of detained persons.

7.2   Advance Directives in the Context of Prison Healthcare

7.2.1   Healthcare-Related Autonomy Rights in Prison

Specific aspects of prison healthcare might affect the way in which advance directives should be used and respected. The first concern regularly raised in the context of prison healthcare relates to limits to free decision-making. Can prisoners be considered truly competent to make their own healthcare decisions? The argument that prisoner-patients do not have the same degree of capacity as patients outside prisons was put forward by legislators in Canton Geneva in the 1980s. The relevant legislation specified that prisoners lack the capacity to freely consent to admission to a psychiatric unit (Elger 2012). Outside the prison context, a distinction is made between voluntary admission to a psychiatric hospital and involuntary commitment. In the former case, a psychiatric patient provides informed consent to being admitted to a psychiatric hospital or unit. In the latter case, the patient is committed to a psychiatric ward against his or her will if predefined conditions apply, such as criteria of dangerousness and an urgent need to treat the person's psychiatric disorder. Involuntary commitment is legally valid only if it is ordered by specifically trained physicians, e.g. chief residents or psychiatrists, who are required to fill in forms that are evaluated by the cantonal body responsible for overseeing involuntary commitments. Before the new cantonal health law came into force in 2006 (Canton de Genève 2006), prisoners could only be admitted to the psychiatric prison ward if the procedure for involuntary commitment was applied, irrespective of the type of admission and of whether the patients concerned provided consent or even requested admission themselves. Only when the new law was prepared did legislators change their position: prisoners are now no longer considered incompetent—on the basis of their legal status—to make decisions regarding admission to a psychiatric hospital. Instead, they now have the legally recognized right to consent to be admitted to a psychiatric ward and, in line with psychiatric admissions outside prisons, a distinction is made between involuntary commitment and voluntary admission.

Similarly to what happened in the past in Geneva, legislators in a number of US states have argued that prisoners cannot meaningfully consent to certain types of health-related interventions. Indeed, in certain US states, research with prisoners is prohibited because legislators doubt that prisoner-patients are truly free and have the capacity to consent (Elger 2008b; Elger and Spaulding 2010). Rather, they might feel under pressure and hence consent to research studies to which they would never have consented outside prison. This could be the case because participation in research may be the only way to access certain types of healthcare in detention—the principle of equivalence is not recognized in the US, and healthcare in prisons may be limited to more or less basic measures (Elger 2008a, 2012).

Closer examination shows that this argument does not mean that prisoners are not able to make rational and reasonable judgments as a result of diminished cognitive abilities or internal "coercion" (due to psychiatric disease). Although the percentage of patients with mental—especially addiction—disorders is higher among prisoners than among the general population, this is not a reason to deny prisoners decision-making capacity in general. From the increased prevalence of psychiatric disease in prisons it follows, rather, that ethically motivated physicians will be particularly careful to assess the competence of each patient.

The questioning of the competence of prisoners in general tends, instead, to be based on the existence of *external* types of coercion in the prison context. Prisoners live in an environment in which they have less autonomy than other individuals making healthcare decisions. As detainees have to follow the orders of prison personnel in most situations, they might conclude that they have to "obey" the physician, too, without resistance or questions. A reasonable detainee might, for example, prefer not to overtly contradict prison physicians even if treatments are not in the prisoner's interest.

However, the proponents of diminished competence of prisoners usually distinguish between different types of medical treatment. In most cases of "ordinary" treatment, prisoners are granted full autonomy rights. It is generally assumed that physicians offer appropriate treatment and that patients are not under undue pressure that might interfere with their decision-making capacity. Only in particular situations such as research and psychiatric inpatient treatment, where abuse and/or stigmatization is feared, would prisoners need to be protected against unfree decisions that are not in their own interest. Legislators in the US and Geneva therefore decided in the past to limit prisoners' autonomy rights in these situations.

The legislative positions described above cannot be justified in accordance with the principle of equivalence. This principle requires that prisoners should not be treated differently than patients outside prison. This implies that, for the individual patient, competence has to be determined for each healthcare decision, and possible pressure or coercion has to be individually evaluated and, if possible, minimized or eliminated. Equivalence means here that external types of pressure or coercion should play the same role in prison healthcare as outside prison. Outside prison, patients in many situations are subject to considerable social pressures, and yet in most cases physicians would not declare all those who have recently been fired incompetent, even if in the first weeks these people have to deal with the pressures of unemployment and financial insecurity. The best way to deal with such external pressures is to try to minimize them. For example, hunger strikes frequently involve groups of politically or religiously motivated prisoners. A certain number of prisoners will continue a hunger strike against their own interests and not on the basis of a freely made decision, but because they face exclusion from their own group and fear possible torture if they give up the strike. If a physician discovers that a patient is continuing a hunger strike only because of group pressure, he or she should try to separate the prisoner from the others—e.g. using the option of admitting the patient to a hospital ward if his or her condition worsens. This will make it easier for the patient to make a decision without direct influence from the group.

Repeated provision of comprehensive information is another way to decrease pressures on free decision-making in the prison context. A physician should explain to prisoners that the principle of equivalence is fully respected, and that the patient has the same autonomy rights as other patients outside prison. The physician could thus correct detainees' false impressions that they are required to obey and carefully evaluate whether patients make decisions in line with their own values. However, physicians can only act in this way if they are granted sufficient independence in the prison healthcare context. They need to have appropriate resources to provide equivalent care, and, in line with the recommendations of the Council of Europe (1998), the hierarchy of health personnel must be separate from the hierarchy of the prison administration and the justice system. This is the case, for example, if the health service is attached to a university hospital or independent public health department, and not to the department of justice and police (Elger 2008a, c, 2011; Sprumont 2012).

We therefore conclude that it is not justified to limit prisoners' autonomy rights concerning healthcare decisions on the grounds of their legal status or the existence of external constraints in prison. Such a position might also have the adverse consequence that any coercive factors in prison which could be corrected or minimized are taken for granted and not changed or adapted. The opposite position—granting prisoners full autonomy rights in health matters—implies an obligation on the part of prison healthcare personnel to closely examine the conditions under which their patients make decisions, and to endeavour to change any factors under their influence that interfere with free decision-making.

7.2.2 Advance Directives in the Context of Imprisonment

Advance directives raise several unique ethical questions in the context of detention. The first relates to where the advance directive has been signed. Advance directives concerning end-of-life care or psychiatric treatment may have been signed before a patient is incarcerated. If the advance directive does not consider the situation of imprisonment, should the directive be regarded as valid if the patient becomes incompetent during imprisonment? Would patients who preferred aggressive chemotherapy outside prison have changed their mind and prefer palliative care if they had known that they would spend the last part of their life in prison? Similarly, if an advance directive concerns authorization of treatment in the event of exacerbation of psychiatric symptoms, would the patient's treatment preferences have varied depending on whether treatment is received inside or outside a prison unit?

The answer to these questions certainly depends on the conditions of detention and healthcare in prisons, which vary widely among countries and settings. It should also be taken into account that, in many situations involving patients who become incompetent during imprisonment, the principle of equivalence would require that the patient be transferred to a prison hospital unit in order to treat the underlying disease. This unit should be specialized to deliver end-of-life and/or psychiatric care. In Switzerland, inpatient units for prisoners exist at two major university hospitals—the Inselspital in Bern and the Hôpitaux Universitaires de Genève (HUG) in Geneva (Elger 2011). Prisoners from all parts of Switzerland will be transferred to these hospitals as soon as they can be transported. In addition, other hospitals closer to the prison where a patient develops acute illness will admit prisoners temporarily for urgent matters. The hospital units in Bern and Geneva are inpatient wards equivalent to other units in these institutions. The main difference is that prison officers are responsible for controlling access and security. This means that prisoner-patients are prevented from leaving the unit, whereas all hospital staff can enter freely, albeit after passing through a metal detector to ensure that weapons or other dangerous materials do not enter the unit. The conditions of hospitalization in one of these prison units are thus not fundamentally different from those in other parts of the university hospitals. One significant difference is that detainee-patients only have access to fresh air for 1 h per day or, in Geneva, not at all, as this unit does not have an outside yard. It is also noteworthy that the HUG prison unit is situated in the basement and the rooms do not have daylight. Furthermore, visits by relatives or friends are generally limited to 1 h per week and need to be approved in advance by the responsible judicial authorities. Although access to treatment might not be different in the prison inpatient units as compared to other parts of the hospitals, the lack of access to fresh air and the restrictions on visitors could clearly impact the quality of life of dying patients and therefore make reasonable patients write different advance directives depending on whether they anticipated being incarcerated or not.

If future imprisonment is considered in an advance directive, the question remains as to whether or not the patient is able to judge the conditions of future imprisonment correctly. It remains unclear whether a person who has never been imprisoned at all, or even prisoners who have not been incarcerated in a particular facility, would ever be able to evaluate correctly in advance how they would feel in that particular future situation. On the other hand, it could be reassuring for psychiatric patients who have never been in prison before to know that their advance directives will be respected even if they are incarcerated, and that they will receive the same treatments in prison as outside. Several means exist to make advance directives more applicable and valid under these conditions. Patients who are at risk for future imprisonment include, in particular, violent patients, drug addicts or previous offenders who have received suspended sentences. These patients should receive specific counselling from healthcare personnel with sufficient knowledge of prison settings. The fact that patients have been informed about the conditions of imprisonment and taken this information into account should be noted in the advance directive. A disadvantage of mentioning future imprisonment in an advance directive arises from the fact that patients carry the directives with them so that they are easily accessible if needed. This means that sensitive information about a future risk of imprisonment is made available to anyone who reads the advance directive. Stigmatization could be a possible consequence if the advance directive is read by other people (e.g. healthcare personnel outside prisons). Patients should be thoroughly informed about the advantages and disadvantages of anticipating incarceration when they write advance directives.

If an advance directive is written by an incarcerated person, it needs to be asked whether or not this person is able to express his or her wishes in an uncoerced way. An experienced independent prison physician who has established a relationship of trust with the patient could judge whether this is the case, and it could be helpful if this physician co-signs the directive and testifies to this fact. However, it must be ensured that future caregivers are able to judge whether the physician's statement is correct. If the physician signing a directive is dependent on and under the influence of prison authorities, this could be an indicator of pressure and would aggravate the risk that advance directives signed in prison do not accurately reflect patients' wishes.

The second main question is whether or not advance directives should be respected in prison. The answer depends, firstly, on the above-mentioned conditions: (1) whether non-incarcerated patients were in a position to anticipate adequately the future prison situation in which their advance directives are supposed to apply and (2) whether it can be assumed that a prisoner writing an advance directive was able to freely express his or her true preferences.

The principle of equivalence indicates that as long as these two conditions are met, advance directives should be respected in prison to the same extent as outside prison. However, studies show that, in many cases, prisoners' advance directives are not respected to the same extent as they would be outside the context of imprisonment (Anno et al. 2004; Linder and Meyers 2007). This may be because the principle of equivalence is not respected in a particular country (e.g. the US). Professionals in the US reported that prison rules limiting or mandating certain treatments would override detainees' advance directives (Scheyett et al. 2010). Other examples related to particular contexts. In France, for example, an advance directive indicating refusal of forced feeding would not be respected in the case of a dying hunger striker because a hunger strike is a treatable condition and a treatment refusal expressed in an advance directive would only be valid in end-of-life situations (such as terminal disease) where death is unavoidable (Dubois et al. 2011). These situations may induce a major conflict for prison physicians forced to decide between respecting professional ethical guidelines—such as the WMA Declarations of Tokyo and Malta (World Medical Association 1975, 1991)—and local requirements of domestic law or prison rules. It remains the role of physicians' professional organizations to address such conflicts, to emphasize the principle of equivalence, and to protect their members if the only appropriate solution is for them to refuse to comply with unethical laws or prison rules.

7.3   Advance Directives in the Context of a Detained Person's Hunger Strike

Publications on advance directives in the context of detention are scarce, and most of them are concerned with the issue of hunger strikes by prisoners (Dubois et al. 2011; Fayeulle et al. 2010; Fessler 2003; Gallot et al. 1964; Gorsane et al. 2007; Oguz and Miles 2005; Stroun 1990; World Medical Association 1991) or detained asylum seekers (Silove et al. 1996; Kenny et al. 2004).

International guidelines stipulate that physicians should respect the autonomous choice of prisoners to die from voluntary fasting. Indeed, Article 6 of the WMA Declaration of Tokyo (World Medical Association 1975) forbids forced feeding of prisoners:

> Where a prisoner refuses nourishment and is considered by the physician as capable of forming an unimpaired and rational judgment concerning the consequences of such a voluntary refusal of nourishment, he or she shall not be fed artificially. The decision as to the capacity of the prisoner to form such a judgment should be confirmed by at least one other independent physician. The consequences of the refusal of nourishment shall be explained by the physician to the prisoner.

Likewise, Article 13 of the WMA Declaration of Malta (World Medical Association 1991) states that "[f]orcible feeding is never ethically acceptable." According to the Preamble to this declaration:

> An ethical dilemma arises when hunger strikers who have apparently issued clear instructions not to be resuscitated reach a stage of cognitive impairment. The principle of beneficence urges physicians to resuscitate them but respect for individual autonomy restrains physicians from intervening when a valid and informed refusal has been made.

An added difficulty arises in custodial settings because it is not always clear whether the hunger striker's advance instructions were made voluntarily and with appropriate information about the consequences.

Article 9 of the guidelines which form part of this declaration affirms that:

Consideration needs to be given to any advance instructions made by the hunger striker. Advance refusals of treatment demand respect if they reflect the voluntary wish of the individual when competent.

Article 9 also requires physicians to evaluate the two conditions discussed above—free, informed decision-making and valid anticipation of the situation to which the advance directive is supposed to apply:

In custodial settings, the possibility of advance instructions having been made under pressure needs to be considered. Where physicians have serious doubts about the individual's intention, any instructions must be treated with great caution. If well informed and voluntarily made, however, advance instructions can only generally be overridden if they become invalid because the situation in which the decision was made has changed radically since the individual lost competence.

If a physician is in doubt about either of these conditions and the patient is incompetent, the guidelines recommend deciding in favour of feeding in order to re-establish competence. However, as stipulated in Article 11, if a hunger striker confirms the treatment refusal, the physician should subsequently comply with the advance directive:

Physicians may consider it justifiable to go against advance instructions refusing treatment because, for example, the refusal is thought to have been made under duress. If, after resuscitation and having regained their mental faculties, hunger strikers continue to reiterate their intention to fast, that decision should be respected. It is ethical to allow a determined hunger striker to die in dignity rather than submit that person to repeated interventions against his or her will.

The World Medical Association's guidelines are internationally recognized documents guiding professional ethics. However, notwithstanding the status of these guidelines, implementation remains a challenge for physicians in various countries. As mentioned above, French authors refer to the fact that hunger strikers' advance directives would not be binding in their country because the patient is not suffering from a terminal disease. Other studies have shown that hunger strikers' advance directives are respected in Canton Geneva and in Dutch and UK prisons (Guilbert 2001; Annas 1995; Fayeulle et al. 2010; Brockman 1999; Fessler 2003; Arda 2002).

A study carried out in France (Fayeulle et al. 2010) found that 7 of 94 physicians who had encountered a hunger strike in prison reported having dealt with advance directives. Of these 7 physicians, 5 indicated that they had respected the advance directives, 1 had not and 1 did not answer this question. The majority of respondents (74 physicians) did not suggest to hunger strikers that they should write advance directives. Of these 74 physicians, 17 said they believed they should employ advance directives in the future, 43 said that advance directives are not of any relevance in a hunger strike situation, and 14 did not respond to this question.

In France, advance directives are conditionally binding. They are to be respected by physicians in cases of terminal illness if the directives correctly predict the situation of care refusal and they have been written by a competent person (Andorno et al. 2009).

Annas (1995) explains the ethical dilemma of physicians who have to judge whether or not the content of an advance directive reflects the true wishes of a detained hunger striker. He points out that hunger striking represents a type of political strategy. Therefore, advance directives may themselves be part of a (manipulative) strategy: the detainee might want to put greater pressure on the authorities by indicating that he or she will continue the strike until death and not accept any kind of resuscitation, although his or her true wish is to continue living.

Others have examined whether starvation itself has an impact on decision-making capacity. As Fessler (2003) notes, in the later stages of a hunger strike, detainees might change their mind concerning resuscitation in the event of loss of consciousness:

> [A]s the possibility of death becomes increasingly real, patients may wish to revise their initial assessments of the value of hunger striking—hypothetical harm contemplated at a distance is not the same thing as real harm staring one in the face, and many would agree that the "reasonable person" may frequently assign greater import to the latter than to the former.

On the other hand, evidence collected by Fessler suggests that the "act of fasting itself leads individuals to both discount the future and be relatively indifferent to harm". If that is the case, he concludes:

> [T]he hunger striker's valuation of the potential costs of fasting may decrease as the fast progresses due not to any lesser affection for life but rather to endogenous changes which reduce the motivational salience of pain, injury, and death.

This might mean that there is no ideal moment of maximum deliberative capacity since at the beginning of a hunger strike detainees might underestimate harms and at a more advanced stage they might become overly indifferent to the dangers because of the endogenous changes induced by the fasting. However, the legal concept of competence does not require a person to reach a maximum level of competence. Since a hunger strike does not cause significant psychiatric disease or cognitive impairment, it would not be in line with the usual definition of legal capacity to declare hunger strikers incompetent or deny them the right to write advance directives. It might still be considered good practice for a physician to ask a detainee who insists on writing an advance directive to do so at a sufficiently early stage of the hunger strike, so as to avoid any doubts about legal competence, though not too early, so as to ensure that the detainee is aware of the reality of future harm.

It remains unclear how a physician can effectively increase the likelihood that an advance directive expresses the hunger striker's true wishes. Hunger strikers seeking to influence the authorities through their behaviour try to convince them that they will persist until death unless the authorities respond positively to their requests. If the authorities know that an advance directive indicates consent to resuscitation, they might not take the hunger striker seriously. This means that a hunger striker wishing to successfully influence the authorities has an interest in writing an advance directive that indicates refusal of resuscitation and feeding even though the detainee might not actually want to die and might secretly ask the prison physician to carry out resuscitation. This would put the physician in a very difficult position, at risk of being accused of failing to respect a written advance directive. Patients have the right to inform others about the content of their advance directives. Therefore, advance directives cannot be kept confidential against the patient's wishes. If physicians and patients agree that confidentiality is to be maintained, the authorities will be uncertain as to whether a detainee is willing to accept death or not and, again, might not take the hunger strike as seriously as if they were aware of an advance directive indicating full treatment refusal. As Silove et al. (1996) observe:

> Physicians in possession of such an advance directive [i.e. one that is kept confidential] may find themselves in the midst of a new game of cat and mouse. Since they would be the only persons other than the hunger strikers who have advance knowledge of the outcome of the strike, such a situation could create further tension between the physicians and the authorities who may use various strategies to gain some indication about the striker's intention from the physician.

These authors also point out that cultural factors further complicate the usefulness and feasibility of advance directives in the context of detention. Many incarcerated persons who start hunger strikes are foreigners, including detained asylum seekers, because a hunger strike may be their only means of securing the authorities' attention. For such detainees, writing an advance directive may not be a realistic option, as they may not understand the legal concept of an advance directive, be able to write, or know the language of the country where they are detained.

Virtually none of the literature on advance directives in the context of hunger strikes discusses the alternative of a surrogate decision maker. We were able to identify only a single reference to this possibility (Oguz and Miles 2005):

> [A] surrogate decision maker may never properly decide to initiate a hunger strike. However, a person with decision making capacity may designate a surrogate decision maker for the time in a hunger strike when he or she loses competence. The designated surrogate decision maker should not be incarcerated or he/she should not be an official of an organisation that is coordinating a collective hunger strike.

While it appears self-evident that the surrogate decision maker should not be from an organization initiating a hunger strike or benefiting from it in any way, it is less obvious why the surrogate cannot be another detainee. The greatest risk is, of course, that a detainee can be put under pressure by the prison authorities, but this might equally be the case for any person within the justice system or the prison hierarchy, and these groups should therefore also be excluded by definition.

With regard to detained hunger strikers, both approaches are somewhat problematic—the designation of a surrogate or advance directives in the form of a living will whereby hunger strikers indicate either their willingness to accept resuscitation and re-feeding or their refusal of such treatment. In the light of these difficulties, the Johannes Wier Foundation for Health and Human Rights (1995; Silove et al. 1996) has suggested an additional safeguard to help ensure respect for the autonomous choice of a hunger striker. The Foundation proposes that hunger strikers appoint a "physician of confidence", whose duty is to represent and defend the autonomy rights and wishes of the hunger striker. The physician of confidence is not exactly the same as a surrogate decision maker, but comes close to the concept in that he or she assumes the role of a patient advocate, ensuring that the patient's wishes are understood and that advance directives are correctly interpreted and respected. Practical problems are that detainees are often not in a position to judge whether or not physicians are independent and impartial. Their ability to make this judgment will depend on the organization of healthcare in each country and on whether or not the independent physician works in the prison. A physician employed by the prison administration could certainly not be considered independent (Kenny et al. 2004). It is also not easy to find a physician who is trusted by both detainees and authorities (Silove et al. 1996). A physician who is not respected by the authorities has limited power to protect the patient's autonomy rights.

Finally, some authors have proposed that such cases should be referred to independent ethics panels or advisory committees that could guide individual physicians and support respect for advance directives in "the highly complex and politicised context in which hunger strikes occur, whether among asylum seekers or other detainees" (Kenny et al. 2004). The advisory committee could be an existing ethics committee of the (independent, non-prison) hospital where a detainee is treated or a committee appointed ad hoc. Its task is to provide advice to the physician on how to manage a detained patient in a hunger strike. Such committees may be helpful in ensuring respect for patients' wishes, but, as Silove et al. (1996) note:

> [There is a risk that they will add] a further layer of complexity to the management of hunger strikes. There can be no guarantee that a committee will make more sensible judgments than the individual physician. Committees may overstep their advisory roles and attempt to insist on compliance with their viewpoints. Disagreement within the committee (or between the committee and the physician) could create yet a further source of pressure for the clinical team.

7.4 Advance Directives in End-of-Life Care for Detained Persons

The number of ageing and diseased prisoners is constantly rising. As a consequence, more and more detainees will die in prison, and many of them will need end-of-life care (Lin and Mathew 2007; Linder and Meyers 2007). Outside prisons, advance directives, including "do not attempt resuscitation" (DNAR) orders, are regularly used in geriatric care and end-of-life decisions (Butterworth 2004; Emanuel 2004). According to the principle of equivalence, prisoners suffering from terminal illness have a right to receive all types of healthcare available to patients outside prison, including palliative care. Providing palliative care to high-security prisoners is difficult, if not impossible, in many countries (Wood 2007). Some countries have opened prison hospices (Boyle 2002; Craig and Craig 1999; Linder et al. 2002) since local community hospices often refuse to admit detainees who are considered dangerous, even if they are terminally ill.

Like any other terminally ill patient, a prisoner facing death in prison might want to refuse care perceived as futile or incompatible with quality of life, wishing to die with dignity. On the other hand, it is well known that certain groups of patients—in the US, black minorities and prisoners in particular—are not worried about receiving too much, but fearful of receiving too little care (Dubler 1998). Studies have shown that detainees have insufficient access to certain life-saving treatments and do not receive adequate pain medication. In particular, opiates, typically used in end-of-life care, are often not prescribed in prisons because of concerns about security and drug dealing (Lin and Mathew 2007; Dubler 1999).

A survey showed that in 2001, 35 of 49 jurisdictions in the US (71 %) permitted inmates to issue advance directives about healthcare. An even higher percentage (86 %, i.e. 42 of 49) allowed DNAR orders (Anno et al. 2004). Linder and Meyers (2007) report that, in spite of state legislation permitting advance directives, patient autonomy in US detention facilities is limited, "as corrections department policies sometimes restrict the use of do-not-resuscitate (DNR) orders and advance medical directives". In addition, research is lacking on whether inmates understand the meaning and scope of advance directives and make use of them (Linder and Meyers 2007; Anno et al. 2004).

Inmates in most countries distrust prison authorities and often also prison physicians. Thus, advance directives can be perceived as a legal means of ensuring respect for patient autonomy and, as such, a trust-enhancing factor in prison healthcare (Anno 2004). While on the one hand care for severely ill prisoners is compromised in many countries by limited access to emergency care facilities and specific medication (Linder and Meyers 2007), respect for advance directives might be impeded by administrative concerns about possible accusations of neglect or indifference. This may lead to overuse of resuscitation and a reluctance to agree to detainees' requests for DNAR orders. Linder and Meyers (2007) describe such a case, where a physician states that:

> [P]risons are likely to offer or provide more aggressive curative attempts (and perhaps even unwanted care) as they seek to dispel any impression of deliberate indifference or withholding treatment from inmates; this can translate into doing "everything possible" to revive a patient in extremis. Thus, the inmate's own wishes may run counter to those of prison administrators and to those of their family.

In order to reduce misunderstandings and increase meaningful patient decision-making (including through advance directives) regarding end-of-life care in prison, it is of crucial importance to grant prison healthcare workers professional independence and access to adequate medical resources. Both are prerequisites to establishing a relationship of trust between physicians and patients, and ensuring that advance directives are used in line with high ethical standards.

7.5 Advance Directives Regulating Psychiatric Treatment of Detained Persons

Studies have shown that psychiatric advance directives increase self-determination of patients with severe mental illness. Advance directives not only help to significantly reduce the number of involuntary commitments to psychiatric units, as shown in a randomized controlled trial (Henderson et al. 2004), but also decrease the frequency of imprisonment for patients diagnosed with bipolar disorder (Quanbeck et al. 2005). While advance directives concerning end-of-life care usually express patients' wishes to limit aggressive treatments that only prolong life at the price of reduced quality of life, psychiatric patients' advance directives often have the opposite aim, i.e. to authorize treatments. Patients with psychosis or bipolar disorder who are in a non-acute phase (and competent) may request to be treated against their will in the event of exacerbation of their psychiatric symptoms. As noted by Scheyett et al. (2007), advance directives can help to ensure that incarcerated patients receive psychiatric treatment faster, without having to await the result of the legal procedures required to obtain authorization for compulsory treatment. These authors carried out a survey among 80 medical and non-medical jail administrators in North Carolina. Many were in favour of advance directives, although they had little ethical or legal knowledge about these instruments and only a few had any practical experience. The arguments put forward against advance directives referred to the limited autonomy rights of incarcerated patients in the US. According to the administrators, individuals' healthcare preferences cannot be respected in the prison context in the same way as outside prison: "in jail we have guidelines we must go by [i.e. internal protocols for management of inmates]". Other respondents explained that "jails would be unable to use psychiatric advance directives because they had limited staffing and resources or because individuals would not be in jail long enough for directives to be utilized". Administrators were in favour of psychiatric advance directives because they helped jail personnel to obtain information they needed on the detainee's background and disease. As one respondent explained:

> [A] lot of times we have people who come in, and we don't know anything about their family or their past, and they can't give us any information [... .] The doctor might prescribe medications that are totally different than what they were taking before [... .] [H]aving psychiatric advance directives would be a good idea.

This statement illustrates several distinctive features of US healthcare, as well as the prison context, and the problems of using advance directives in these settings. Since access to basic mental healthcare is not granted through social insurance in the US, marginalized populations and the uninsured, in particular, are undertreated. A significant number of psychiatric patients end up in prison because it is the only place where they will receive at least some treatment. Advance directives in this context become a form of medical records, explaining patients' disease, wishes and social context. Since they are carried by the patient him/herself, the confidentiality of this sensitive medical information is sacrificed. The acceptance of decreased confidentiality among medical and non-medical prison administrators is not surprising, as the US is known to apply different standards of medical ethics to prisoner-patients than to non-incarcerated patients (Elger 2008a), including with respect to confidentiality. The principle of equivalence would, however, require that advance directives remain confidential and accessible only to medical personnel bound by professional confidentiality obligations. Not only have psychiatric patients a right to privacy, both in and outside prison, but also, from a consequentialist viewpoint, confidentiality ensures that patients can speak more confidently to their caregivers, without fearing that revealing information about their disease will lead to stigmatization or other significant adverse consequences, including compromised employment chances in the future. For mentally ill prisoners, confidentiality is particularly important since they are at risk for abuse. Vulnerable groups, including psychiatric patients, are more often victims of violence, exploitation and rape than non-psychiatric patients (Dumond 2003; Robertson 2003). When questioned about psychiatric advance directives in a study using mixed methods, students with severe mental illnesses (SMIs) considered psychiatric advance directives (PAD) "potentially problematic because PAD use raised a risk of breaching student privacy and stigmatizing students with SMIs" (Scheyett and Rooks 2012).

7.6 Conclusions

In countries where prison healthcare is truly independent and the principle of equivalence is fully respected, advance directives are a means of increasing patients' self-determination. Problems persist in prisons, similar to those existing outside the prison context, as regards a patient's ability to anticipate the medical and other details of a future situation in which he or she becomes incompetent. Some doubts remain about thresholds and the evaluation of competence and as to whether decisions are made freely without undue pressures. Informed consent of competent patients is a prerequisite for healthcare both in and outside prison. The main ethical issue for prison healthcare personnel is the need for careful evaluation of possible pressures that exist in the prison context and might lead to discrepancies between patients' true wishes and their written advance directives. In difficult situations, inside and outside prisons, physicians should ask for a second opinion, provided by a physician from a different institution (World Medical Association 1991). As shown by the complexities of hunger strikes, the maintenance of confidentiality is important in order to permit patients to trust their physicians, communicate their wishes fully and change their mind when their condition worsens. To avoid advance directives becoming a publicity tool rather than an expression of patients' true wishes, we have in the past advised detainees on hunger strike against writing an advance directive that would not have been truly authentic.

If the principle of equivalence is fully respected, prisoners should have access to palliative care and be allowed to die with dignity. This implies that, towards the end of life, the ethically appropriate solution is not only to allow and respect advance directives, but to permit timely compassionate release of dying prisoners from detention, as stipulated by international soft law (Council of Europe 1998).

The situation changes if the principle of equivalence is not recognized—as is the case explicitly in the US, for example, but also implicitly in many other countries, despite the fact that many of them have ratified relevant human rights conventions (Council of Europe 1950, 1987). In these places, advance directives as a means of consenting in advance to psychiatric care can help to ensure that imprisoned patients with severe mental illness receive more timely and appropriate care, but they will also entail a loss of confidentiality. In addition, owing to the unique challenges in the correctional environment, including the risk of coercion, ill prisoners "may also experience pressure to select treatments desirable to the prison staff" (Thomas and Watson 1998).

Advance directives used to refuse treatments might be perceived as even more delicate in detention facilities in these countries, since they might result from patients' frustration with poor conditions and lack of respect for the principle of equivalence and therefore not convey the wishes patients would have expressed outside of prisons. Although such decisions taking into account limited quality of life in prison may be considered rational, they are ethically problematic if the reasons have to do with violations of prisoners' basic human rights (e.g. lack of access to healthcare, and inhuman or degrading conditions of detention). Furthermore, since refusal of life-saving treatment is irreversible if death is the consequence, it is even more important to be sure that the wishes expressed are authentic and not the result of undue pressures or unacceptable prison conditions. Physicians in such settings find themselves in very difficult and ethically intractable situations. They need strong support from professional medical associations. Preventive measures should comprise nationally and internationally enforced guidelines (Swiss Academy of Medical Sciences 2002), as well as collective refusal, supported by these associations, to provide medical care under unethical circumstances.

Although advance directives are tools supporting self-determination and autonomy, their use in prisons poses unique challenges and requires further research.

References

Andorno, R., N. Biller-Andorno, and S. Brauer. 2009. Advance health care directives: Towards a coordinated European policy? *European Journal of Health Law* 16(3): 207–227.

Annas, G.J. 1995. Hunger strikes: Can the Dutch teach us anything? *BMJ* 311: 1114–1115.

Anno, B.J. 2004. Prison health services: An overview. *Journal of Correctional Health Care* 10(3): 287–301.

Anno, B.J., C. Graham, J.E. Lawrence, and R. Shansky. 2004. *Addressing the needs of elderly, chronically ill, and terminally ill inmates*. Middletown: Criminal Justice Institute.

Arda, B. 2002. How should physicians approach a hunger strike? *Bulletin of Medical Ethics* 181: 13–18.

Boyle, B.A. 2002. The Maryland Division of Correction hospice program. *Journal of Palliative Medicine* 5(5): 671–675.

Brockman, B. 1999. Food refusal in prisoners: A communication or a method of self-killing? The role of the psychiatrist and resulting ethical challenges. *Journal of Medical Ethics* 25(6): 451–456.

Butterworth, A.M. 2004. Advance directives. Vital to quality care for elderly patients. *Advance for Nurse Practitioners* 12(3): 69–75.

Canton de Genève. 2006. K103. Loi sur la santé du 7 avril 2006.

Council of Europe. 1950. *The European convention on human rights.* http://www.hri.org/docs/ ECHR50.html. Accessed 20 June 2010.

Council of Europe. 1987. *European convention for the prevention of torture and inhuman or degrading treatment or punishment.* http://conventions.coe.int/Treaty/en/Treaties/Html/126. htm. Accessed 20 June 2012.

Council of Europe. 1998. *Committee of ministers rec no R (98) 7 concerning the ethical and organisational aspects of health care in prison.* www.coe.ba/pdf/Recommendation_No_R_ 98_7_eng.doc. Accessed Apr 2011.

Cox, D.M., and G.A. Sachs. 1994. Advance directives and the patient self-determination act. *Clinics in Geriatric Medicine* 10(3): 431–443.

CPT. 2006. European Committee for the Prevention of Torture and Inhuman or Degrading Treatment or Punishment (CPT). The CPT standards: "Substantive" sections of the CPT's general reports. http://www.cpt.coe.int/en/documents/eng-standards-scr.pdf. Accessed Feb 2013.

Craig, E.L., and R.E. Craig. 1999. Prison hospice: An unlikely success. *The American Journal of Hospice & Palliative Care* 16(6): 725–729.

Dubler, N.N. 1998. The collision of confinement and care: End-of-life care in prisons and jails. *The Journal of Law, Medicine & Ethics* 26(2): 149–156.

Dubler, N.N. 1999. Commentary: Promoting ethical flexibility. *Journal of Pain and Symptom Management* 17(2): 145–146.

Dubois, F., E. Sudre, A. Porte, R. Bedry, and S. Gromb. 2011. Evolution and follow-up of hunger strikers: Experience from an interregional hospital secured unit. *La Revue de Médecine Interne* 32(11): 669–677. doi:10.1016/j.revmed.2011.05.001.

Dumond, R.W. 2003. Confronting America's most ignored crime problem: The prison rape elimination act of 2003. *The Journal of the American Academy of Psychiatry and the Law* 31(3): 354–360.

Elger, B.S. 2008a. Medical ethics in correctional healthcare: An international comparison of guidelines. *The Journal of Clinical Ethics* 19(3): 234–248; discussion 254–259.

Elger, B.S. 2008b. Research involving prisoners: Consensus and controversies in international and European regulations. *Bioethics* 22(4): 224–238. doi:10.1111/j.1467-8519.2008.00634.x.

Elger, B.S. 2008c. Towards equivalent health care of prisoners: European soft law and public health policy in Geneva. *Journal of Public Health Policy* 29(2): 192–206. doi:10.1057/jphp. 2008.6.

Elger, B.S. 2011. Prison medicine, public health policy and ethics: The Geneva experience. *Swiss Medical Weekly* 141: w13273. doi:10.4414/smw.2011.13273.

Elger, B.S. 2012. Autonomie und Macht – Medizinethik im Gefängnis. *Studia Philosophica* 70(2012): 163–185.

Elger, B.S., and A. Spaulding. 2010. Research on prisoners – A comparison between the IOM Committee recommendations (2006) and European regulations. *Bioethics* 24(1): 1–13. doi:10.1111/j.1467-8519.2009.01776.x.

Emanuel, L.L. 2004. Advance directives and advancing age. *Journal of the American Geriatrics Society* 52(4): 641–642. doi:10.1111/j.1532-5415.2004.52177.x.

Emanuel, L.L. 2008. Advance directives. *Annual Review of Medicine* 59: 187–198. doi:10.1146/ annurev.med.58.072905.062804.

European Court of Human Rights. 1994. Case of Hurtado v. Switzerland. http://hudoc.echr.coe.int/sites/eng/pages/search.aspx?i=001-57868

Fayeulle, S., F. Renou, E. Protais, V. Hedouin, G. Wartel, and J.L. Yvin. 2010. Management of the hunger strike in prison. *La Presse Médicale* 39(10): e217–e222. doi:10.1016/j.lpm.2010.01.012.

Fessler, D.M. 2003. The implications of starvation induced psychological changes for the ethical treatment of hunger strikers. *Journal of Medical Ethics* 29(4): 243–247.

Gallot, H.M., A. De Mijolla, and S. Schaub. 1964. Remarks on a case of hunger strike. *Annales de Médecine Légale, Criminologie, Police Scientifique et Toxicologie* 44: 354–356.

Goffin, T. 2012. Advance directives as an instrument in an ageing Europe. *European Journal of Health Law* 19(2): 121–140.

Gorsane, I., K. Zouaghi, R. Goucha, F. El Younsi, H. Hedri, S. Barbouch, T. Ben Abdallah, F. Ben Moussa, H. Ben Maiz, and A. Kheder. 2007. Acute renal failure in a prisoner after hunger strike. *La Tunisie Médicale* 85(3): 234–236.

Guilbert, P. 2001. Le jeûne de protestation en médecine pénitentiaire. Épidémiologie genevoise (prison préventive et quartier cellulaire hospitalier) et analyse de la prise en charge en Suisse et dans les pays européens. Thèse: Med: Université de Genève (220 pages). http://archive-ouverte.unige.ch/vital/access/manager/Repository/unige:108. Accessed 2 July 2012.

Henderson, C., C. Flood, M. Leese, G. Thornicroft, K. Sutherby, and G. Szmukler. 2004. Effect of joint crisis plans on use of compulsory treatment in psychiatry: Single blind randomised controlled trial. *BMJ* 329(7458): 136. doi:10.1136/bmj.38155.585046.63.

Johannes Wier Foundation for Health and Human Rights. 1995. *Assistance in hunger strikes: A manual for physicians and other health personnel in dealing with hunger strikers*. Amersfoort: Johannes Wier Foundation for Health and Human Rights.

Kenny, M.A., D.M. Silove, and Z. Steel. 2004. Legal and ethical implications of medically enforced feeding of detained asylum seekers on hunger strike. *The Medical Journal of Australia* 180(5): 237–240.

Lin, J., and P. Mathew. 2007. Prison inmates and palliative care. *JAMA: The Journal of the American Medical Association* 298(21): 2481; author reply 2481. doi:10.1001/jama.298.21.2481-a.

Linder, J.F., and F.J. Meyers. 2007. Palliative care for prison inmates: "Don't let me die in prison". *JAMA: The Journal of the American Medical Association* 298(8): 894–901. doi:10.1001/jama.298.8.894.

Linder, J.F., S.R. Enders, E. Craig, J. Richardson, and F.J. Meyers. 2002. Hospice care for the incarcerated in the United States: An introduction. *Journal of Palliative Medicine* 5(4): 549–552. doi:10.1089/109662102760269788.

Monahan, J., M. Swartz, and R.J. Bonnie. 2003. Mandated treatment in the community for people with mental disorders. *Health Affairs (Millwood)* 22(5): 28–38.

Oguz, N.Y., and S.H. Miles. 2005. The physician and prison hunger strikes: Reflecting on the experience in Turkey. *Journal of Medical Ethics* 31(3): 169–172. doi:10.1136/jme.2004.006973.

Quanbeck, C.D., B.E. McDermott, and M.A. Frye. 2005. Clinical and legal characteristics of inmates with bipolar disorder. *Current Psychiatry Reports* 7(6): 478–484.

Robertson, J.E. 2003. Rape among incarcerated men: Sex, coercion and STDs. *AIDS Patient Care and STDs* 17(8): 423–430. doi:10.1089/108729103322277448.

Scheyett, A.M., and A. Rooks. 2012. University students' views on the utility of psychiatric advance directives. *Journal of American College Health* 60(1): 90–93. doi:10.1080/07448481.2011.572326.

Scheyett, A.M., M.M. Kim, J.W. Swanson, and M.S. Swartz. 2007. Psychiatric advance directives: A tool for consumer empowerment and recovery. *Psychiatric Rehabilitation Journal* 31(1): 70–75.

Scheyett, A.M., J.S. Vaughn, and A.M. Francis. 2010. Jail administrators' perceptions of the use of psychiatric advance directives in jails. *Psychiatric Services* 61(4): 409–411. doi:10.1176/appi. ps.61.4.409.

Sebo, P., P. Guilbert, B. Elger, and D. Bertrand. 2004. Le jeûne de protestation: un défi inhabituel pour le médecin. *Médecine & Hygiène* 2508 (08/12/2004): 54–62.

Silove, D., J. Curtis, C. Mason, and R. Becker. 1996. Ethical considerations in the management of asylum seekers on hunger strike. *JAMA: The Journal of the American Medical Association* 276(5): 410–415.

Sprumont, D. 2012. The independence of medicine in prison: A small epilogue of the case Rappaz. *Revue Médicale Suisse* 8(332): 607–609.

Stroun, J. 1990. Hunger strike in a prison environment. *Revue Médicale de la Suisse Romande* 110(5): 451–456.

Swiss Academy of Medical Sciences. 2002. *The exercise of medical activities in respect of detained persons.* http://www.samw.ch/en/Ethics/Guidelines/Currently-valid-guidelines.html. Accessed 20 June 2012.

Swiss Confederation. 2012. *Swiss civil code.* http://www.admin.ch/ch/e/rs/2/210.en.pdf. Accessed Feb 2013.

Thomas, D.L., and J.M. Watson. 1998. Advance directives in a correctional setting. *Psychology, Public Policy, and Law* 4(3): 878–899.

United Nations. 1982. Principles of medical ethics relevant to the role of health personnel, particularly physicians, in the protection of prisoners and detainees against torture and other cruel, inhuman or degrading treatment or punishment. Adopted by General Assembly resolution 37/194 of 18 Dec 1982. http://www2.ohchr.org/english/law/medicalethics.htm

United Nations. 1990. Basic principles for the treatment of prisoners. Adopted and proclaimed by General Assembly resolution 45/111 of 14 Dec 1990. http://www2.ohchr.org/english/law/basicprinciples.htm

Wood, F.J. 2007. The challenge of providing palliative care to terminally ill prison inmates in the UK. *International Journal of Palliative Nursing* 13(3): 131–135.

World Medical Association. 1975. WMA declaration of Tokyo – Guidelines for physicians concerning torture and other cruel, inhuman or degrading treatment or punishment in relation to detention and imprisonment (Revised 2006). http://www.wma.net/en/30publications/10policies/c18/. Accessed 3 July 2012.

World Medical Association. 1991. WMA declaration of Malta on hunger strikers (Revised 2006). http://www.wma.net/en/30publications/10policies/h31/. Accessed 3 July 2012.

Part III
Effects on Family, Friends and Professional Relations

Chapter 8
Advance Directives and the Physician-Patient Relationship: A Surprising Metamorphosis

Mark P. Aulisio

8.1 Introduction

In Chap. 1 of this volume, the history of what might be termed the advance directive (AD) movement in Western medicine is clearly and carefully traced. Though stretching back to at least the late 1960s, this history famously includes the well-known and controversial US court cases of Karen Quinlan (*In re Quinlan* 1976) and Nancy Cruzan (1990). Both cases involved "permanently unconscious" patients (to use current terminology) who could have no direct conscious interaction with physicians, caregivers, family or friends. Despite this, or more accurately and paradoxically *because* of this, Karen Quinlan and Nancy Cruzan have dramatically impacted each of these relationships—none more so than our focus, the physician-patient relationship—by giving powerful motive force to the AD movement, first in the US and then wherever in the world Western medicine is practised (Schiff et al. 2006; Rodriguez-Arias et al. 2007; Baumann et al. 2009; Kranidiotis et al. 2010; Schaden et al. 2010; van Wijmen et al. 2010; Schuklenk et al. 2011; Ting and Mok 2011; Kim et al. 2012).

Indeed, it was the public outcry in the US during and after the Cruzan case that directly led to the passage of the federal Patient Self-Determination Act (PSDA 1990), which required that every adult patient in US healthcare institutions receiving federal support be asked upon admission whether he or she had an advance directive and, if not, be offered the opportunity to complete one. Now, more than 20 years after the PSDA was passed with great promise and high expectations, there is a growing chorus of critics who think that the AD movement has failed and that the whole project is deeply flawed (Fagerlin and Schneider 2004; Moseley et al. 2005; Collins et al. 2006; Perkins 2007; Castillo et al. 2011). In the eyes of

M.P. Aulisio, Ph.D. (✉)
Center for Biomedical Ethics, MetroHealth Medical Center, Cleveland, OH, USA

Department of Bioethics, Case Western Reserve University, Cleveland, OH, USA
e-mail: mark.aulisio@case.edu

P. Lack et al. (eds.), *Advance Directives*, International Library of Ethics, Law, and the New Medicine 54, DOI 10.1007/978-94-007-7377-6_8,
© Springer Science+Business Media Dordrecht 2014

the critics, there is precious little to show for the huge effort invested over the last 30-plus years. Whether or not the movement has failed, however, depends on what it was supposed to achieve, and this in turn is not just a matter of its proponents' expectations. It is also, rather, a matter of the normative justification for the movement in the first place. Herein I consider (1) the normative justification for the AD movement and, in light of that, its ultimate goal; (2) how the movement has failed to achieve this goal; and, lastly, (3) the impact of the movement on the physician-patient relationship. While acknowledging the understandable disappointment of the AD movement's most ardent supporters and the well-aimed critiques of many of its detractors, I contend that the movement itself has been a startling success in its impact on the physician-patient relationship—serving as a catalyst for a dramatic metamorphosis.

8.2 Normative Justification for the Advance Directive Movement

A variety of justifications for advance directives can be found in the literature. Nearly all appeal to autonomy as a moral value tied to conceptions of moral agency in which *agency* in the relevant sense is tied to *personhood*. Advance directives are straightforwardly taken to be vehicles for making effective the autonomous moral agency of persons in circumstances in which such agency could not otherwise be effective because the person in question, the patient, lacks the capacity to exercise that agency (Buchanan and Brock 1989; Ikonomidis and Singer 1999). Most often, of course, these are circumstances, such as the prospective withdrawal of ventilator support, for which the autonomous moral agency of patients would be highly relevant and in which the decisions would normally be made by a decisionally capacitated patient. I will refer to this as the *standard justification* for advance directives.

Given this justification, then, it is not surprising that philosophical critiques of advance directives often focus on their operative or implied conceptions of moral agency and personhood. For example, advance directives have been criticized as fundamentally flawed because they presume continuity of personal identity in certain categories of patients, such as the permanently unconscious or severely demented, for whom continuity of personal identity is in doubt—some of the very patients for whom advance directives would potentially be most helpful (Buchanan 1988). Advance directives have also been attacked as little more than modern-day Ulysses contracts or a form of selling oneself into slavery and, therefore, deeply incompatible with autonomous moral agency (May 1997; Tollefsen 1998; Varelius 2011). Most severely, the general conception of moral agency that is supposed to underlie advance directives has itself been excoriated as an utterly impoverished form of social atomism—"rugged individualism"—that is little more than the fancy of moral philosophers and at odds with the highly social and communal manner in which humans actually make decisions (Ikonomidis and Singer 1999).

Whatever its philosophical merit, the standard justification for advance directives implies a rather clear and practical goal for the AD movement, that is, that the movement ideally result in all adults (with capacity) having an advance directive which both indicates their wishes *and* designates a surrogate to make healthcare decisions when they are unable to do so. This goal would seem to be supported by the requirements of the Patient Self-Determination Act which, as mentioned above, was adopted in response to public outcry in the US during and after the Cruzan case. The PSDA requires that healthcare organizations receiving Medicare or Medicaid funds:

1. Give [patients] at the time of admission a written summary of:

   - [their] health care decision-making rights (Each state has developed such a summary for hospitals, nursing homes, and home health agencies to use.)
   - the facility's policies with respect to recognizing advance directives.

2. Ask [patients] if [they] have an advance directive, and document that fact in [the patient's] medical record if [they] do. (It is up to [patients] to make sure they get a copy of it).
3. Educate their staff and community about advance directives.
4. Never discriminate against patients based on whether or not they have an advance directive. Thus, it is against the law for them to require either that [a patient] have or not have an advance directive. (See: www.americanbar.org/groups/public_education/resources/law_issues_for_consumers/patient_self_determination_act.html; wording modified for number and person agreement.)

Though the PSDA had its early critics (Wolf et al. 1991), its passage was heralded as a major step forward for patients, enabling them to exert their moral autonomy over decisions even when they were unable to actually make those decisions for themselves (La Puma et al. 1991). Nowhere was this more pressing, as Quinlan and Cruzan poignantly illustrated, than at the end of life, when one might be imprisoned indefinitely in a dying process prolonged by unwanted medical interventions with no hope for meaningful recovery (Pence 2008). Advance directives would be the mechanism enabling patients to avoid this sad state of affairs—extending control over decisions where little to none previously existed.

Given the exigency of the issue, it is not surprising that PSDA-like legislation has been passed in Canada, the European Union and Japan with, at least implicitly, a fundamental normative goal consonant with that of the standard justification: that all adults (with decisional capacity) have an advance directive which both indicates their wishes *and* designates a surrogate to make healthcare decisions when they are unable to do so. With this goal in mind, then, how has the AD movement fared?

8.3 Has the Advance Directive Movement Achieved Its Goal?

Setting aside the deep philosophical questions that critics have aimed at the standard justification for advance directives as discussed above, it is fair to ask— more than 20 years after the passage of the PSDA and more than 35 years post-Quinlan—how well or poorly the AD movement has fared in achieving its goal.

One might expect the AD movement to have made the greatest progress in the US, both because the movement started here and also because American culture ostensibly places the greatest emphasis on autonomy. However, the empirical data, albeit not comprehensive, is far from encouraging. For example, 85–95 % of adult patients in the US do not have an advance directive of any kind (Kirschner 2005; Jones et al. 2011). Of those who claim to have an advance directive, anyone who has worked in healthcare knows that even fewer actually produce them. More alarming still, significant percentages of patients in high-risk populations who would be expected to have advance directives do not have them. For example, one study showed that only 41 % of patients with chronic heart failure had an advance directive of any kind (Dunlay et al. 2012).

A slightly more favourable study, carried out by the National Center for Health Statistics of the US Department of Health and Human Services Centers for Disease Control and Prevention, found that 28 % of home healthcare patients, 65 % of nursing home residents and 88 % of discharged hospice patients had some type of advance directive (Jones et al. 2011). A closer analysis is, unfortunately, less auspicious because the study covered more than six types of advance directives, including living will, DNR order, do not hospitalize, feeding restrictions, medical restrictions and organ donation. When broken down by category, it turned out that the DNR orders, in particular, skewed the data set, with only 17 % of home healthcare patients, 18 % of nursing home residents and 26 % of discharged hospice patients having living wills. For nursing home residents, living wills appear to have been broadly construed to include healthcare proxy or surrogate directives; the same does not appear to be true for home healthcare and discharged hospice patients. When the latter are included, an additional 5 % of home healthcare patients have healthcare power of attorney or proxy/surrogate directives, while an additional 16 % of discharged hospice patients have these. However, even with these factored in, only 22 % of home healthcare and 42 % of discharged hospice patients have either a living will or a proxy/surrogate directive—an astoundingly low number considering the patient populations.

Though these numbers have led more than a few commentators to deem the AD movement an abject failure, things may be worse still for proponents of the movement. The study of heart failure patients cited above found that, of the 41 % who had an advance directive, many failed to specify their wishes regarding standard types of end-of-life care, such as cardiopulmonary resuscitation, mechanical ventilation or medically administered ("artificial") nutrition and hydration (Dunlay et al. 2012).

This in itself may not be so damning, since surrogate decision-makers, designated or not, have long been touted as the key to overcoming the obvious shortcomings that notoriously complicated the interpretation of early living wills (Fagerlin and Schneider 2004). Unfortunately, empirical data on surrogate decision-making suggests that surrogates have problems of their own. For example, a comprehensive literature review and meta-analysis (Shalowitz et al. 2006) concluded that:

> On average, patient-designated and next-of-kin surrogates incorrectly predict patients'
> end-of-life treatment preferences in one third of cases. Also, it appears that the 2 most
> commonly endorsed methods for improving surrogate accuracy—patient designation of a
> surrogate and prior discussion of treatment preferences with surrogates—are not effective.

The latter point, i.e. that patient designation of a surrogate and prior discussion of treatment preferences with surrogates doesn't seem to make any significant difference, is especially concerning, as it suggests that efforts (such as "National Decision Day") to get people to do just these things may be misguided. Sadly, this pessimistic conclusion comports with the findings of the famous Study to Understand Prognoses and Preferences for Outcomes and Risks of Treatments (SUPPORT 1995), namely, that enhancing opportunities for physician-patient communication didn't seem to result in a positive impact on patient outcomes at the end of life.

These apparently devastating failures have led some to argue that it is time to pull the plug on the whole AD movement. Advance directives, on this view, cannot be redeemed by tinkering around the edges. The problem is that the concept itself is, much like the naive conception of agency it presupposes, hopelessly impoverished. The advance directive project, so the criticism goes, is a futile attempt to control the uncontrollable, i.e. the circumstances in which death comes to each of us and how we shall respond to those circumstances. As one commentator (Perkins 2007) puts it:

> Many experts blame problems with completion and implementation, but the advance
> directive concept itself may be fundamentally flawed. Advance directives simply presuppose
> more control over future care than is realistic. Medical crises cannot be predicted in
> detail, making most prior instructions difficult to adapt, irrelevant, or even misleading.

Indeed, the situation with advance directives and end-of-life care is so fraught with peril and futility that this commentator is led to appeal to no less than Albert Camus in an attempt to make this otherwise dark night just a bit less dark:

> At the end of Camus' *The Plague*, the main character—a physician named Rieux—reflects
> on his role throughout the plague epidemic. He realizes that, along with providing care that
> had to be given "by all who . . . strive . . . to be healers," he bore witness to patients'
> suffering [. . .]. Physicians surely have the duty to fight disease in most circumstances, but
> physicians always have the still greater duty to see patients and survivors through their
> suffering and thereby to bear witness to it. Perhaps that greater duty lifts medicine from a
> mere occupation to a true profession.

The AD movement has perhaps, on this view, forgotten that medicine is not a "mere occupation"—the physician not merely a technician who mechanically applies and follows the instructions of an abstract, ill-suited and poorly conceived document that is naively supposed to guide, by autonomy's bright light, care in a patient's last days. If this devastating critique is well aimed, and it certainly appears to be, it is clear not only that the AD movement is and has been an abject failure, but also that it has necessarily been such—a sadly ironic exercise in futility from the start.

8.4 Physician-Patient Relationship and Advance Directives: Catalyst for Metamorphosis?

Though the AD movement clearly has its shortcomings and, even more clearly, has failed to achieve the straightforward goal of every adult patient having an advance directive, I want to suggest that the movement has not been an abject failure. More strongly, I want to argue that, when measured against a more fundamental norma- tive goal than was articulated above, the AD movement has been a rather startling success. I will make this argument by offering an alternative justification for advance directives and, by implication, an alternative normative goal for the AD movement. This justification is more concerned with power and authority than control, and is more social and political than it is moral, even if moral justifications might successfully be offered. Furthermore, this justification does not presuppose an impoverished, atomistic conception of human nature, as does, arguably, the standard justification considered above. To get at this justification, a brief excursus into the history of medical practice is required.

Historically, physicians in the West have had a social status akin to clergy. This stems from multiple sources, including the close affinity between "healing" and religious practices. Priest, rabbi, imam, shaman, witch doctor all, like the medical doctor, have "healing" as part of their art and craft. Similarly, the Hippocratic Corpus appears to have religious origins and characterizes what might be called the special vocation or calling of the doctor in noble and religious terms. For example, the earliest Greek versions of the Hippocratic Oath seem to be closely tied to what Al Jonsen (2000) called "dogmas of the Pythagorean faith", after followers of "the philosopher, mystic and mathematician Pythagoras". Central to the Oath, and running through the entire Hippocratic Corpus, is a constellation of notions related to physicians doing "good" or "helping", and "abstaining from intentional injustice or harm" (Jonsen 2000). On the interpretation of Jonsen and others, doing "good" meant (at least in part) recognizing the limits of their craft in treating only those who can be helped and not those whose condition is beyond the reach of the craft, as they may be "harmed" by such treatment and, therefore, wronged ("injustice") as well (Jonsen 2000).

Not surprisingly, those in need of the healing power of the priest, rabbi, imam, shaman, witch doctor, or, for our purposes, the physician, are largely vulnerable and passive. Patients, as the word's origin (Latin *pati* "to suffer") suggests, are characterized in largely passive terms. Patients, qua those who are suffering, are *acted upon* by the art and craft of the physician. To the extent that patients are active at all, it is only insofar as they must, as Hippocrates' *Epidemics* I enjoins, "cooper- ate with the physician" as he practises his art and craft (Jonsen 2000).

The religious and mystical view of the physician, along with the concomitant view of the passive and compliant patient, endured through the Roman period in the work of Galen, who averred that the "best doctor is also a philosopher" and sees in the ideal physician a paragon of virtue who will "practice temperance and despise money: all evil actions that men undertake are done either at the prompting of greed or under the spell of pleasure" (Jonsen 2000). A good doctor is a "*good* doctor" where "good" just means something like holy, virtuous and wise—a robust, substantive notion of "good". The robust and socially powerful "good doctor" theme can be found through the medieval period in the works of the revered Persian Islamic scholar Avicenna and in those of the equally revered Jewish scholar Maimonides, down to the present day through Cabot, Osler, Leake and others (Jonsen 2000).

The notion of the "good doctor" is so deeply ingrained in the tradition of Western medicine that even today one occasionally hears it used. Contemporary works of medical ethics such as Howard Brody's *The Healer's Power* (1992) and Margaret Mohrmann's *Medicine as Ministry* (1995) critically draw on these themes as well. While Brody eloquently and rightly connects "healing" and "power" and Mohrmann insightfully sees medicine as a form of ministry, popular myths about the capabilities of modern medical science have arguably combined with the August tradition of Western medicine to create a perhaps historically unparalleled social standing and power for physicians. It is always perilous to guess where one might be situated on a historical arc, but our best guess is that this arc reached its zenith, at least in the US, sometime during the second half of the twentieth century, with the advent of what seemed to be almost god-like capabilities to cure the sick and extend lives—antibiotics, vaccinations, transplants, dialysis machines and respirators.

Advancement, progress and power, however, also brought fear and trepidation. From the heinous abuses of the Nazi doctors, to the US military's experimentation on its own soldiers, to Belding Scribner's pejoratively termed Seattle "God Committee", to the infamous US Public Health Service/CDC Tuskegee syphilis study, the potential abuse of the power of modern medicine became increasingly evident (Jonsen 1993, 1997; Pence 2008). These abuses highlighted the need for protections of human research subjects—de facto limitations on the power of medical researchers—but also began to erode the notion of "good doctor" in the popular imagination.

Outside of the research setting, the legalization of abortion in the US (1973), Karen Quinlan's very public tragedy (1976), the Baby Doe controversies (1982–1984), and eventually the emotionally charged plight of Nancy Cruzan (1990) further eroded the "good" in "good doctor" in the minds of many. Thus, whether in research or clinical practice, even as modern medicine made great advances in healing and helping those who suffer, it also risked harming them and making them suffer even more—further victimizing those who were already victimized by disease—and creating a need for protections for patients and limitations on physicians' power.

The kinds of cases discussed above, in both research and clinical medicine, led not only to the recognition that either might be used for good or evil, but also over time to two much deeper realizations as well. First, the most central disputes in all of the above cases are not fundamentally about the medical or scientific facts, but about the normative ethical and legal issues. This is especially true for Quinlan and Cruzan, the two cases most relevant for our purposes. Should Karen Quinlan's parents or the care team at St. Clare's Hospital be allowed to make the decision about whether or not to turn off her respirator? If one or the other should be allowed to make the decision, are there some bases for such a decision that should be viewed as legitimate and others that should be viewed as illegitimate? Similarly, who, if anyone, should be allowed to decide whether to discontinue the medical provision of nutrition and hydration to Nancy Cruzan and on what bases? There are, of course, no strictly medical or scientific answers to these questions. The questions are fraught with ethical and legal values and depend on the relevant ethical and legal frameworks under discussion. In short, medical decision-making in Quinlan, Cruzan and all of the above cases is fundamentally *value-laden*. Just how value-laden medical decision-making can be is illustrated by the following simple hypothetical case (Aulisio et al. 2003):

The Case of the Multitalented Motorcyclist: "A medical no-brainer?"
A patient is brought into the ER as a result of a bad motorcycle accident. Due to lack of blood flow, the patient's right leg is in danger. A vein could be taken from the wrist and placed in the leg to save it, leaving the patient with a minor disability in the wrist. The patient insists on having the leg amputated.

In this case, which is deliberately thin on, and detached from, medical facts, it turns out that the patient has not suffered a head injury, nor is he psychotic or under the influence of mind-altering substances. Indeed, ex hypothesi, the patient is alert and oriented to person, place and time and has a firm grasp of the options and their consequences. The patient adamantly insisted that his wrist be left alone and that his leg be amputated if necessary. The case is rigged, of course, and the patient turns out to be a famous and accomplished virtuoso pianist. What appears to be a medically obvious decision—a medical "no-brainer"—turns out to be value dependent; this raises the question of whose values should drive decision-making or, to put it differently, of who should be allowed to make this value-laden decision (Aulisio et al. 2000, 2003).

In addition to the *fact* that medical decision-making is value-laden, the very public and controversial cases discussed above (and others) led to a growing realization that we don't all share the same values, even where research and clinical medicine are concerned. Karen Quinlan's parents were at odds with members of the care team and hospital administration at St. Clare's Hospital; Nancy Cruzan's parents with the State of Missouri. As these cases aptly demonstrate, value-laden issues being addressed by individuals and communities with different values can lead to uncertainty and even bitter conflict. This again calls for an answer to the question of whose values should drive decision-making, or who should be allowed to make the decision.

The increasingly pluralistic context in which Western medicine is practised reflects the deep reality that only a very thin notion of the "good" may be widely shared in post-modern society. In the context of the Greek city-state, with its shared notion of the good and medicine's very limited ability to "help and not harm", the judgment of the "good doctor" may have been enough—not so for our post-modern context. As the world in which Western medicine is practised has dramatically expanded, the publicly shared notion of the good that crosses that world's nations, communities and individuals has proportionately contracted, with important implications for medical practice. The "good doctor" may not adequately share the values of the patient to be relied on in making a "good" decision *for that patient*. This, in part, helps to account for the rise of informed choice (whether consent or refusal) as central to research participation or clinical medical decision-making with adults. Whatever one thinks of the importance of the *moral* value of autonomy and the underlying conception of *moral agency*, one might still recognize the importance of autonomy as a *political* value that responds to the increasingly splintered notion of the good and the need, therefore, to limit physicians' power—the power of the "good doctor"—in light of that.

So what is the relevance of this for advance directives and the AD movement? The "good doctor" at the zenith of social standing and power during the latter half of the twentieth century was confronted by a growing recognition of the value-laden nature of research and clinical medicine and the ever-more pluralistic context in which that research and clinical medicine was conducted. This created a need for a countervailing power—a countervailing *political* power—in medical decision-making, which emerged in the rise of informed choice with its undergirding political value, autonomy. The alternative justification for advance directives and the AD movement arises out of this context. Advance directives are justified as extensions of the countervailing *political* value of autonomy—a limitation of the authority and standing of the "good doctor"—into the realm in which patients are most vulnerable to the abuse of physicians' power, i.e. when they are unable to make decisions for themselves or even to participate in decision-making.

8.5 Concluding Remarks

At the outset of this chapter, I articulated what was termed the "standard justification" for advance directives, i.e. that advance directives are straightforwardly taken to be vehicles for making effective the autonomous moral agency of persons in circumstances in which such agency could not otherwise be effective because the person in question, the patient, lacks the capacity to exercise that agency. This justification is subject, as noted, to certain types of philosophical criticisms, not the least of which is that it presupposes a dubious conception of moral agency, a sort of social atomism or rugged individualism, that is patently at odds with the highly social and communal manner in which humans actually make decisions. In addition, the normative goal of the standard justification was articulated as that of having all adults with decisional capacity fill out an advance directive which both indicates their wishes *and* designates a surrogate to make healthcare decisions when they are unable to do so. As we have seen, the AD movement appears to have been a rather abject failure in light of this goal. What of the alternative justification?

First, the alternative justification for advance directives in no way presupposes social atomism or rugged individualism. Recall that, on the alternative justification, advance directives are conceived as extensions of the countervailing *political* value of autonomy—a limitation of the authority and standing of the "good doctor"—into the realm in which patients are most vulnerable to the abuse of physicians' power, i.e., when they are unable to make decisions for themselves or even to participate in decision-making. Because the operative conception of autonomy is *political* rather than moral, the concern is more with shifting the focus of power and authority away from the "good doctor" and towards the patient than it is with assessing whether advance directives are authentic expressions of the moral agency of the patient (person). Furthermore, contrary to what the standard justification's commitment to social atomism would appear to allow, the alternative justification of advance directives grounded in autonomy as a countervailing *political* (rather than moral) value leaves room for patients to engage—either directly or through their advance directives—in highly social and communal decision-making, so long as patients remain the source of political authority for such decision-making.

Second, the alternative justification provides the basis for a different and more fundamental normative goal of the AD movement than that of the standard justification. On the standard justification, the normative goal of the AD movement was articulated as that of having all adults with decisional capacity fill out an advance directive which both indicates their wishes *and* designates a surrogate to make healthcare decisions when they are unable to do so. As we have seen, not only do the numbers of adult patients with advance directives in the US alone fall woefully short of this goal, there are other problems with advance directives that make them miserable failures if they are supposed to be authentic extensions of the moral agency of patients. The alternative justification for advance directives as articulated in this section, however, does not require that they be authentic expressions of moral agency. It requires, rather, that they be ways to politically limit physicians' power, shifting it away from physicians and towards patients. Furthermore, with respect to sheer numbers, the fact that any significant percentage of the adult population in the US has an advance directive is itself remarkable, given that we are *only* 30 years on from the PSDA. Lastly and most importantly, the fact that nearly all of the literature *critiquing* advance directives is focused on how well or poorly they ensure respect for *patient* values in decision-making is itself an indication of the startling success of the AD movement: it demonstrates a dramatic transformation—even metamorphosis—of the physician-patient relationship. Over the span of a few decades, Western medicine has moved from a "good doctor" or physician-centred paradigm of medical decision-making to one centrally concerned with patient values and how well or poorly these are respected in decision-making. The AD movement was both a product of and a major contributor to this shift and, as such, it is not much of an overstatement to call it a revolution.

Acknowledgments I would like to thank David Essi, MA, for his work in formatting this chapter and, along with David, Nicole Deming, JD, MA, Jason Gatliff, Ph.D., and Monica Gerrek, Ph.D., for good conversation that helped shape the contents of this chapter.

References

Aulisio, M.P., R.M. Arnold, and S.J. Youngner. 2000. Health care ethics consultation: Nature, goals, and competencies. A position paper from the Society for Health and Human Values-Society for Bioethics Consultation Task Force on Standards for Bioethics Consultation. *Annals of Internal Medicine* 133(1): 59–69.

Aulisio, M.P., R.M. Arnold, and S.J. Youngner (eds.). 2003. *Ethics consultation: From theory to practice*. Baltimore: Johns Hopkins University Press.

Baumann, A., G. Audibert, F. Claudot, and L. Puybasset. 2009. Ethics review: End of life legislation – The French model. *Critical Care* 13(1): 204.

Brody, H. 1992. *The healer's power*. New Haven: Yale University Press.

Buchanan, A. 1988. Advance directives and the personal identity problem. *Philosophy & Public Affairs* 17(4): 277–302.

Buchanan, A.E., and D.W. Brock. 1989. *Deciding for others: The ethics of surrogate decision making*. Cambridge/New York: Cambridge University Press.

Castillo, L.S., B.A. Williams, S.M. Hooper, C.P. Sabatino, L.A. Weithorn, and R.L. Sudore. 2011. Lost in translation: The unintended consequences of advance directive law on clinical care. *Annals of Internal Medicine* 154(2): 121–128.

Collins, L.G., S.M. Parks, and L. Winter. 2006. The state of advance care planning: One decade after SUPPORT. *The American Journal of Hospice & Palliative Care* 23(5): 378–384.

Cruzan v. Director, Missouri Dept. of Health, 497 U.S. 261 (1990)

Dunlay, S.M., K.M. Swetz, P.S. Mueller, and V.L. Roger. 2012. Advance directives in community patients with heart failure. *Circulation. Cardiovascular Quality and Outcomes* 5(3): 283–289.

Fagerlin, A., and C.E. Schneider. 2004. Enough. The failure of the living will. *The Hastings Center Report* 34(2): 30–42.

Ikonomidis, S., and P.A. Singer. 1999. Autonomy, liberalism and advance care planning. *Journal of Medical Ethics* 25(6): 522–527.

In re Quinlan, 355 A.2d 647, *cert. denied sub nom. Garger v. New Jersey*, 429 U.S. 922 (1976).

Jones, A.L., A.J. Moss, and L.D. Harris-Kojetin. 2011. Use of advance directives in long-term care populations. *NCHS Data Brief* 54: 1–8.

Jonsen, A.R. 1993. The birth of bioethics. *The Hastings Center Report* 23(6): S1–S4.

Jonsen, A.R. 1997. The birth of bioethics: The origins and evolution of a demi-discipline. *Medical Humanities Review* 11(1): 9–21.

Jonsen, A.R. 2000. *A short history of medical ethics*. New York: Oxford University Press.

Kim, S.S., W.H. Lee, J. Cheon, J.E. Lee, K. Yeo, and J. Lee. 2012. Preferences for advance directives in Korea. *Nursing Research and Practice* 2012: 873892.

Kirschner, K.L. 2005. When written advance directives are not enough. *Clinics in Geriatric Medicine* 21(1): 193–209.

Kranidiotis, G., V. Gerovasili, A. Tasoulis, E. Tripodaki, I. Vasileiadis, E. Magira, V. Markaki, C. Routsi, A. Prekates, T. Kyprianou, P.M. Clouva-Molyvdas, G. Georgiadis, I. Floros, A. Karabinis, and S. Nanas. 2010. End-of-life decisions in Greek intensive care units: A multicenter cohort study. *Critical Care* 14(6): R228.

La Puma, J., D. Orentlicher, and R.J. Moss. 1991. Advance directives on admission. Clinical implications and analysis of the Patient Self-Determination Act of 1990. *JAMA: The Journal of the American Medical Association* 266(3): 402–405.

May, T. 1997. Reassessing the reliability of advance directives. *Cambridge Quarterly of Healthcare Ethics* 6(03): 325–338.

Mohrmann, M.E. 1995. *Medicine as ministry: Reflections on suffering, ethics, and hope.* Cleveland: Pilgrim Press.

Moseley, R., A. Dobalian, and R. Hatch. 2005. The problem with advance directives: Maybe it is the medium, not the message. *Archives of Gerontology and Geriatrics* 41(2): 211–219.

Patient Self-Determination Act 1990 42 U.S.C. 1395 cc (a).

Pence, G.E. 2008. *Classic cases in medical ethics: Accounts of the cases and issues that define medical ethics.* New York: McGraw-Hill Higher Education.

Perkins, H.S. 2007. Controlling death: The false promise of advance directives. *Annals of Internal Medicine* 147(1): 51–57.

Rodriguez-Arias, D., G. Moutel, M.P. Aulisio, A. Salfati, J.C. Coffin, J.L. Rodriguez-Arias, L. Calvo, and C. Herve. 2007. Advance directives and the family: French and American perspectives. *Clinical Ethics* 2(3): 139–145.

Schaden, E., P. Herczeg, S. Hacker, A. Schopper, and C.G. Krenn. 2010. The role of advance directives in end-of-life decisions in Austria: Survey of intensive care physicians. *BMC Medical Ethics* 11: 19.

Schiff, R., P. Sacares, J. Snook, C. Rajkumar, and C.J. Bulpitt. 2006. Living wills and the Mental Capacity Act: A postal questionnaire survey of UK geriatricians. *Age and Ageing* 35(2): 116–121.

Schuklenk, U., J.J. van Delden, J. Downie, S.A. McLean, R. Upshur, and D. Weinstock. 2011. End-of-life decision-making in Canada: The report by the Royal Society of Canada expert panel on end-of-life decision-making. *Bioethics* 25(Suppl 1): 1–73.

Shalowitz, D.I., E. Garrett-Mayer, and D. Wendler. 2006. The accuracy of surrogate decision makers: A systematic review. *Archives of Internal Medicine* 166(5): 493–497.

SUPPORT. 1995. A controlled trial to improve care for seriously ill hospitalized patients. The study to understand prognoses and preferences for outcomes and risks of treatments (SUPPORT). The SUPPORT Principal Investigators. *JAMA: The Journal of the American Medical Association* 274(20): 1591–1598.

Ting, F.H., and E. Mok. 2011. Advance directives and life-sustaining treatment: Attitudes of Hong Kong Chinese elders with chronic disease. *Hong Kong Medical Journal* 17(2): 105–111.

Tollefsen, C. 1998. Response to "Reassessing the Reliability of Advance Directives" by Thomas May (CQ Vol. 6, No. 5). *Cambridge Quarterly of Healthcare Ethics* 7(04): 405–413.

van Wijmen, M.P., M.L. Rurup, H.R. Pasman, P.J. Kaspers, and B.D. Onwuteaka-Philipsen. 2010. Design of the advance directives cohort: A study of end-of-life decision-making focusing on advance directives. *BMC Public Health* 10: 166.

Varelius, J. 2011. Respect for autonomy, advance directives, and minimally conscious state. *Bioethics* 25(9): 505–515.

Wolf, S.M., P. Boyle, D. Callahan, J.J. Fins, B.N. Jennings, J.L. Nelson, J. Barondess, A.D.W. Brock, R. Dresser, L. Emanuel, S. Johnson, J. Lantos, D.R. Mason, M. Mezey, D. Orentlicher, and F. Rouse. 1991. Sources of concern about the Patient Self-Determination Act. *New England Journal of Medicine* 325(23): 1666–1671.

Chapter 9
Advance Directives and the Role of Family and Close Persons – Legal Provisions and Challenges

Margot Michel

9.1 Introduction

In recent decades, respect for patient autonomy has become one of the core principles of modern Western medical law and ethics. It calls for essentially unconditional recognition of an individual's right to self-determination. Self-determination in a legal sense encompasses the individual's right to decide, of his or her own free will, on medical interventions. It is based on international and national law and is usually derived from or seen as a part of the human right to physical and mental integrity and dignity. Therefore, the right to self-determination is itself regarded as a human right. The right to self-determination includes the right to give or withhold consent to a medical intervention, even if this intervention would be necessary to keep the patient concerned alive. Thus, it also allows for a right to die, at least insofar as death will occur naturally if no life-sustaining treatment is provided.

As a basic principle, only the patient's consent to treatment renders medical interventions lawful. Article 5 of the Oviedo Convention (Council of Europe 1997a, b; for detailed discussion see Sect. 9.2.1) stipulates that an intervention in the health field may only be carried out after the person concerned has given free and informed consent to it. Medical interventions which are against the competent patient's expressed wishes or to which consent has not been given are seen, from a legal viewpoint, as a violation of the patient's physical integrity. Here, the term "intervention" is to be understood in its widest sense, covering not only treatment but also interventions performed for the purpose of preventive care, diagnosis, rehabilitation or research (Council of Europe 1997b: Explanatory Report paragraph 34). Therefore, the right to

M. Michel (✉)
Assistant Professor for Private Law, Institute of Law, University of Berne, Berne Switzerland,
e-mail: margot.michel@ziv.unibe.ch

P. Lack et al. (eds.), *Advance Directives*, International Library of Ethics, Law, and the New Medicine 54, DOI 10.1007/978-94-007-7377-6_9,
© Springer Science+Business Media Dordrecht 2014

self-determination in the health field is legally exercised by giving or withholding consent.

It is worth clarifying that the right to self-determination does not require a decision which is reasonable from an objective point of view; rather, a competent person is free to make a choice in accordance with his or her own beliefs and values, which need not correspond to prevailing societal values. As the right to self-determination is so far-reaching—encompassing even the right not to avert one's own death—enjoyment of this right depends on the presence of certain capabilities in the decision-maker.

Firstly, the person has to be competent to make this kind of decision. What this actually means has to be determined in detail under national law, but legal competence usually calls for the capacity to understand the decision in question and its consequences, and to form one's own free will on the basis of the facts and of one's own values and beliefs. Only persons who have these capabilities are permitted and required to bear the legal and practical consequences of their decisions. Therefore, persons who lack these capabilities also lack the right to self-determination and cannot validly consent to or refuse treatment.

Secondly, the consent has to be free and informed. Apart from the fact that patients must not be put under any pressure, they have to be fully informed about the intervention being contemplated, including its purpose, nature, consequences, risks and possible alternatives, in a manner they can understand (Council of Europe 1997b:Explanatory Report paragraphs 35 and 36).

As incompetent patients lack legal decision-making capacity, some form of surrogate decision-making is required. With regard to the above-mentioned function of consent as justifying the violation of physical integrity that is involved in any medical intervention, there are two main questions to be answered from a legal viewpoint:

1. Who is entitled or bound to make a decision on behalf of the incompetent patient, and what are the guiding principles for surrogate decision-making?
2. Can a person issue legally binding directives in anticipation of becoming incompetent at a later time—for example, by writing a living will?

Basically, there are two different forms of surrogate decision-making—extension of self-determination and delegation of decision-making power to a third party. Although they are usually treated as separate issues, they can be combined: the option of creating a durable power of attorney in a living will allows for an autonomous decision with regard to designation of a proxy while at the same time leaving the detailed content of decision-making open and adaptable to specific future circumstances and developments (for the different forms of advance directives see Andorno 2010:119).

In this chapter, I will explore the intersections between advance directives and surrogate decision-making by family members, close persons or proxies. For though advance directives, especially in the form of a living will, are intended to extend self-determination to a time when the patient is no longer competent and therefore by definition place a strong emphasis on autonomy, the exercise of this autonomy essentially depends on the assistance of third parties, as shown by Brauer (2008). Brauer further argues that not only the implementation of a living will but also—in a fundamental way—the understanding and interpretation of directives in specific circumstances is dependent on the cooperation of third parties. In view of the dependency between the patient and those who interpret advance directives, she then identifies a form of relational autonomy as the underlying concept of living wills (Brauer 2008:232–233)—and therefore rejects the notion that living wills are based on a purely individualistic concept of autonomy.

The interpretation of living wills is also very important from a legal viewpoint. Family members and close persons, as well as physicians, play an important role in the interpretation of advance directives, e.g. when the wording is not entirely clear and easily applicable to the concrete situation. Correct interpretation becomes absolutely crucial when conflicts arise between the patient's expressed or presumed wishes and his or her interests.

On the other hand, delegation of decision-making power to a proxy, family member or close person does not necessarily deprive the patient of self-determination. By implementing specific guidelines for surrogate decision-making by proxies, the law can ensure that—as far as possible—surrogate decisions are aligned with the patient's wishes and views and are not merely paternalistic. Moreover, the law can and must provide measures against the risk of family members or close persons failing to act in the patient's interests.

Family members and close persons thus play an important role when decisions have to be taken for incompetent patients. The question I will address in this chapter is whether the law grants power of representation to family members and close persons and, if so, what guiding principles are implemented for surrogate decision-making. I will also examine whether and how the power of representation is limited, and what procedural safeguards can and should be implemented to protect the incompetent patient's interests.

Firstly, I will take an in-depth look at the European legal framework relating to advance directives and the role assigned to the family and close persons of incompetent patients. Here, my discussion will be confined to the treaties and soft law of the Council of Europe. I will then consider the moral authority of the family and possible conflicts. Finally, I will present and analyse the most relevant provisions of the new Swiss adult protection law.

9.2 Legal Framework Concerning the Role of the Family

9.2.1 Convention on Human Rights and Biomedicine

The Council of Europe's Convention for the Protection of Human Rights and Dignity of the Human Being with regard to the Application of Biology and Medicine (Council of Europe 1997a, b), signed in Oviedo on 4th April 1997 (referred to hereinafter as the Oviedo Convention), entered into force on 1st December 1999, having been ratified by five countries (including four member states of the Council of Europe). To date, it has been ratified by 29 states and signed by 6 others. However, the Oviedo Convention has not been signed by Germany or the United Kingdom, among other member states of the Council of Europe. It has been ratified by Switzerland, where it came into effect on 1st November 2008.

The Oviedo Convention is one of the Council of Europe's most important conventions, setting out for the first time legally binding standards for the sphere of biomedicine on an international level. It is designed as a core convention and therefore only sets out the most important principles regarding the protection of human rights and fundamental freedoms in the application of biomedicine; more detailed provisions are included in additional protocols to the convention, dealing with specific issues such as research, cloning, transplantation or genetic testing. In particular, the aim of the convention is to guarantee everyone's rights and fundamental freedoms and, in particular, their integrity and to secure the dignity and identity of human beings in this sphere (Council of Europe 1997b:Explanatory Report paragraph 17; see Andorno 2011:75). Under Article 1, each party is required to take in its internal law the necessary measures to give effect to the provisions of the convention. This means that each state that has ratified the Oviedo Convention has to ensure that its domestic law conforms to the Convention's provisions. This can be done either by enacting or revising the relevant national legislation; alternatively, in states (e.g. Switzerland) that fall under the system of direct applicability, provisions of the convention that qualify as *self-executing* can be directly applied. A provision qualifies as self-executing when it is drafted sufficiently clearly to be directly applied and therefore can be invoked before court without prior implementation in national law. In particular, this is the case when a provision formulates individual rights (Council of Europe 1997b:Explanatory Report paragraph 20). According to the rules of international law, the provisions of an international convention have to be interpreted autonomously. Thus, the meaning of a provision has to be found by exploring its phrasing, sense, purpose and aim in the context of the convention itself (cf. Article 31 of the Vienna Convention on the Law of Treaties: "A treaty shall be interpreted in good faith in accordance with the ordinary meaning to be given to the terms of the treaty in their context and in the light of its object and purpose."; see Herdegen 2012:336).

The Oviedo Convention sets out a *minimal standard* regarding the protection of human rights and dignity in the application of biology and medicine. While ratifying states are free to provide for more extensive protection of human rights and dignity in certain fields, they must not fall short of the standards set out in the Oviedo Convention.

Two provisions of the Oviedo Convention are of particular relevance to the topic of this chapter—Article 6, which deals with the protection of persons who are not able to consent, and Article 9, which concerns previously expressed wishes.

Article 6(3) states that:

> Where, according to law, an adult does not have the capacity to consent to an intervention because of a mental disability, a disease or for similar reasons, the intervention may only be carried out with the authorization of his or her representative or an authority or a person or body provided for by law.

This provision specifies that it has to be determined under national law whether or not a person is capable of validly consenting to a medical intervention. If he or she lacks this capacity, the convention states—in Article 6(1)—that interventions may only be carried out for his or her own benefit (the "principle of protection"), except in the case of medical research or removal of regenerative tissue, for which special provisions apply. Article 6(3) includes a very important specification: medical interventions carried out for the benefit of an incompetent patient require the authorization of the legal representative or an authority, person or body provided for by law. Except in the case of emergency situations (Article 8), this precludes interventions being legitimated by the physician who performs them. States that ratified the Oviedo Convention but lacked legal provisions regarding the appointment of a healthcare proxy for an incapacitated adult therefore needed to revise their legislation. Switzerland, for example, had to revise its adult protection law, which now specifies an order of precedence for close persons entitled to give consent to a medical intervention (see Sect. 9.4). The patient's representative or the authority has to be fully informed about the medical intervention in order to be able to validly give or withhold consent (Article 6(4)). According to Article 6(3), the incompetent patient must be given the opportunity to *participate* in the decision-making process as far as possible. Therefore, national law must provide for the wishes of an incompetent patient being taken into account. Nevertheless, it has to be considered that only competent persons are permitted and required to bear the consequences of their decisions, especially when these decisions would be harmful to them. Thus, the extent to which incompetent patients' wishes can be taken into account in the authorization procedure is limited by these patients' interests. At the very least, however, the significance and circumstances of the intervention have to be explained to the incompetent patient in a manner he or she is able to understand, and his or her opinion has to be obtained and considered (Council of Europe 1997b:Explanatory Report paragraph 46).

Article 9 of the Oviedo Convention deals with the previously expressed wishes of patients who are no longer in a state to express their wishes at the time of a medical intervention: previously expressed wishes "shall be taken into account" (for a discussion of Article 9 and its ambiguity see Andorno 2011:75–78). The phrase "previously expressed wishes" is not further elucidated in the Explanatory Report to the Oviedo Convention. According to the principles of public international law, the terms of an international treaty—as mentioned above—have to be interpreted autonomously, i.e. within the context of the treaty itself, taking into consideration its aim and spirit. When understood in a broad sense, for which I argue here, the phrase "previously expressed wishes" can encompass living wills as well as wishes that were only expressed verbally (e.g. when talking to relatives, friends or physicians). This interpretation of Article 9 is underpinned by the Explanatory Report, which states (in paragraph 62) that the obligation to take previously expressed wishes into account "does not mean that they should necessarily be followed", and by Parliamentary Assembly Resolution 1859 (Council of Europe 2012), which (in paragraph 3) applies the term "previously expressed wishes" to "advance directives, living wills or continuing powers of attorney", while at the same time stating (in paragraph 7.2) that advance directives should, *in principle*, be made in writing. When a relatively weak legal force is accorded to previously expressed wishes, as is the case in the Oviedo Convention, there is no reason why they should be restricted to written living wills. Moreover, neither the phrasing of the English nor the French version of the Convention indicates that a written living will is required.

9.2.2 Convention for the Protection of Human Rights and Fundamental Freedoms

The Council of Europe's Convention for the Protection of Human Rights and Fundamental Freedoms, signed on 4th November 1950, contains several provisions that are self-executing in states falling under the system of direct applicability. In these states, the provisions in question are directly binding for legislators, courts and authorities. The individual person can invoke the relevant rights before national courts. The central supervisory organ for the Convention for the Protection of Human Rights and Fundamental Freedoms is the European Court of Human Rights in Strasbourg.

The Convention seeks to ensure the individual's right to life (Article 2) and protection from inhuman or degrading treatment (Article 3). The latter provision encompasses fundamental protection of human dignity and physical and mental integrity. The provision is applicable in cases of major violations of physical or mental integrity that are an expression of disrespect for human dignity. However, according to the case law of the European Court of Human Rights, the right to self-determination, as well as the right to physical and mental integrity, is predominantly protected by Article 8 of the Convention, which concerns the right to respect for private and family life (Grabenwarter and Pabel 2012:230). It is from this provision that the right to personal autonomy and the principle of consent are derived. Interventions, even minor ones, are not allowed without the consent of the person affected. Treatment refusal by competent adults must be respected; there must be no compulsory treatment against the will of a competent adult (Council of Europe 2012: paragraph 1; Frowein and Peukert 2009).

As the Oviedo Convention fleshes out the Convention for the Protection of Human Rights and Fundamental Freedoms for the sphere of biomedicine, it provides more concrete guidance on questions concerning the right to self-determination and the protection of incompetent patients. However, in states that have not signed or ratified the Oviedo Convention, the Convention for the Protection of Human Rights and Fundamental Freedoms still provides a very important legal basis for patients' rights.

9.2.3 Recommendations and Resolutions of the Council of Europe

In addition to conventions, which have to be ratified by member states and then become legally binding, the Council of Europe has also issued several recommendations and resolutions relevant to advance directives and the role of the family and close persons. As recommendations and resolutions do not require ratification by member states, they are not legally binding per se. Nevertheless, as soft law, they do significantly influence national legislation, as well as the case law of the European Court of Human Rights, and therefore have to be taken into account. For the purposes of this chapter, I will present them in chronological order.

9.2.3.1 Principles Concerning the Legal Protection of Incapable Adults

Committee of Ministers Recommendation No. R(99)4 (Council of Europe 1999a) deals with principles concerning the legal protection of incapable adults. Having regard to the principles of subsidiarity and proportionality, states should, according to this recommendation, provide a range of measures of protection, making it possible to take different degrees of incapacity and various situations into account (Principle 2.1). It should therefore be possible for protection measures which are only needed for a short time, as well as minor and routine decisions relating to healthcare, to be taken by proxies deriving their powers directly from the law (Principles 2.5 and 2.8). It is thus not necessary for public authorities to be routinely involved. According to Principle 3.1, national law should recognize that different degrees of incapacity may exist and that incapacity may vary from time to time; therefore, a measure of protection should not result automatically in a complete removal of legal capacity. Moreover, formal measures of protection should only be taken when less formal arrangements or assistance provided by family members or other persons are not feasible or sufficient (Principle 5.2). Incapable persons have the right to participate in the decision-making process: they should be adequately informed and given the opportunity to express their views (Principle 9.3). In addition, national law should limit the power of representatives and determine which decisions they cannot take (Principle 19.1). As persons with mental disorders are particularly vulnerable, it is recommended that these patients should only be treated without their consent if the disorder is of a serious nature and serious harm to their health would result if they were left untreated (Principle 25).

9.2.3.2 Protection of the Human Rights and Dignity of the Terminally Ill and the Dying

Parliamentary Assembly Recommendation 1418 (Council of Europe 1999b) contains important clarifications of the previously adopted Oviedo Convention with regard to terminally ill and dying patients and is therefore also relevant to advance directives. The recommendation stresses that national laws should provide protection not only against prolongation of the dying process of a person against his or her will (paragraph 8 ii), but also against limitation of life-sustaining treatment due to economic reasons (paragraph 8 v). This latter issue can be expected to assume particular relevance in future, as financial resources in healthcare become more and more limited and patients and their family members nowadays often request more extensive treatment than physicians would recommend. The will of a terminally ill or dying person to refuse treatment is to be respected, but there must be safeguards to ensure that such wishes are not formed under economic pressure or under the influence of another person (paragraph 9 b iii).

Member states are to define criteria of validity for advance directives, especially with regard to the scope of instructions given in advance and the limits of the power of appointed proxies. Surrogate decisions taken by proxies have to be clearly connected to previous, freely expressed statements of the patient and must not be based solely on general value judgments present in society. In case of doubt, the decision must always be for life and the prolongation of life (paragraph 9 b iv)—a principle which could also prove to be challenging in view of limited resources and the increasing expectations of patients and their families. When no advance directive or living will exists, the patient's right to life must not be infringed upon (paragraph 9 b vi). Provided they do not violate human dignity, the expressed wishes of a terminally ill or dying person with regard to particular forms of treatment are to be taken into account (paragraph 9 b v). This allows for non-compliance with advance directives in situations where human dignity is jeopardized—for example, when a patient previously refused pain medication but is now, when incapable, in a state of intolerable pain. Member states are to define a catalogue of treatments that must not be withheld or withdrawn under any conditions (paragraph 9 b vi). However, the ultimate therapeutic responsibility rests with the physician (paragraph 9 b v). It is to be recognized that the terminally ill or dying person's wish to die never constitutes any legal claim to die at the hand of another person (paragraph 9 c ii), nor does it constitute a legal justification to carry out actions intended to bring about death (paragraph 9 c iii).

9.2.3.3 Principles Concerning Continuing Powers of Attorney and Advance Directives for Incapacity

Recommendation CM/Rec(2009)11 (Council of Europe 2009) aims to promote self-determination by means of continuing powers of attorney and advance directives (Principle 1.1)—the latter term encompassing both legally binding and merely advisory advance directives (Andorno 2010:123). With regard to the threshold nature of legal capacity, special attention is to be paid to the question of how incapacity should be determined and what kind of evidence should be required by law (Principle 7.2). While a continuing power of attorney must be in writing (Principle 5.1), the capable granter should be free to revoke it at any time (Principle 6).

The recommendation sets out guidelines for decision-making by the attorney: he or she has to act in accordance with the continuing power of attorney and in the interests of the granter (Principle 10.1). Moreover, he or she has to inform and consult the granter and take the latter's past and present wishes and feelings into account and give them due respect (Principle 10.2). Therefore, the right of the patient to participate in the decision-making process has to be respected, notwithstanding his or her incapacity. Additionally, states are to provide a system of supervision (Principle 12.2) and regulate conflicts between the granter's and the attorney's interests (Principle 11).

9.2.3.4 Protecting Human Rights and Dignity by Taking into Account Previously Expressed Wishes of Patients

The most recent resolution (Council of Europe 2012), based on the assumption that advance directives, living wills and continuing powers of attorney are very important means of protecting the human rights and dignity of incompetent patients, considers it to be essential that member states enact and fully implement legislation on advance directives (paragraphs 3, 4, 6). In addition to the principles already enshrined in the Oviedo Convention, it is recommended that member states should promote advance directives, living wills and/or continuing powers of attorney. Self-determination in these forms is to be given priority over other measures of protection (paragraph 7.1). In principle, living wills, advance directives and/or continuing powers of attorney should be made in writing and be fully taken into account when properly validated and registered (paragraph 7.2). In addition, national law should provide for the possibility of a public appointment in cases where the individual has made no appointment him or herself, provided this is in the best interest of the individual (paragraph 7.3). If prior instructions contained in advance directives and/or living wills are against the law, or good practice or do not correspond to the actual situation in which the author now finds him- or herself, they should not be applied (paragraph 7.4). Further, complicated forms or expensive formalities are to be avoided, as these render advance directives less accessible to all (paragraph 7.5). Advance directives should be reviewed at regular intervals (the Assembly recommends once a year) and revocable at any time (paragraph 7.6). To combat abuse, the member states are to establish a system of supervision and empower a competent authority to investigate and intervene, in particular if an attorney is not acting in accordance with the continuing power of attorney or in the interests of the granter (paragraph 7.7).

9.3   Moral Authority of the Family and Possible Conflicts

The moral authority of family members who are to be involved in decision-making on behalf of an incompetent patient is usually derived from their assumed close relationship (see, for example, the rationale for the new Swiss adult protection law given in the Federal Council's Dispatch: Schweizerischer Bundesrat 2006:7014). It is also common for the moral authority of family members appointed as healthcare proxies to be justified on the basis of traditions or societal agreements (Zellweger et al. 2008:201). In addition, as pointed out in feminist literature, a common Western understanding of autonomy fails to adequately take account of the fact that it actually depends on relationships with other people and can only be developed in the first place with the help of family members or close persons. Thus, social relationships are not only a prerequisite for the development of autonomy but also an ongoing condition of its implementation (Zellweger et al. 2008:201).

Giving decision-making power to relatives can certainly be to the best advantage of the incompetent patient. Family members—at least close ones—are usually familiar with the patient's values, preferences and background. Moreover, the informal appointment of family members as healthcare proxies is much faster and easier than involving a public authority, especially when surrogate decision-making is only needed temporarily (Gevers et al. 2012:65). This corresponds to the principle of proportionality, according to which public intervention in individuals' lives should only occur if no less invasive measure is possible. It has also been argued that Article 8 of the European Convention on Human Rights and Fundamental Freedoms, which protects family life as a fundamental right, "supports the idea that close relatives should be allowed to act as informal representatives, unless a court, or the patient himself, has decided otherwise" (Gevers et al. 2012:65).

However, the routine appointment of family members as healthcare proxies involves certain drawbacks and risks. Not all family members are close to each other, and in dysfunctional families, or cases where there is a history of family mistrust or conflict, this approach would certainly be detrimental. Besides, it has to be taken into consideration that family members and close persons have their own interests, which may conflict with those of the patient. For example, family members may be influenced by the fear of having to bear the costs of care-giving or the burdens of home care. Conversely, family members may find it difficult to detach themselves from the patient, and the fear of losing him or her can make it emotionally impossible to decide in accordance with the patient's wishes.

If family members are designated as healthcare proxies in quite an informal way (as is the case in Switzerland), there is often no review of their capacity and suitability for making decisions of this kind; moreover, the question of supervision arises—how can it be ensured that family members do in fact act in the patient's interests? And how can an informal representative be removed from his or her position when not acting in the patient's interests (Jox et al. 2008:62)? In this context, it is worth noting that the withdrawal of power of attorney from family members and its transfer to public authorities was originally intended to better protect incompetent persons from arbitrary decisions of family members. This concern has to be carefully addressed by legislation, as it is by no means groundless: as recently as 2008, the European Court of Human Rights decided a case where the mother of a young man suffering from mental problems had him committed to a mental hospital and declared incompetent. She was appointed as his guardian, and the young man was as a result being isolated from the outside world (ECHR, *Case of Shtukaturov v. Russia*, Application no. 44009/05, 27 March 2008).

Furthermore, empirical studies have shown that healthcare proxies often do not correctly understand the patient's presumed wishes (Zellweger et al. 2008:210). Determining what an incompetent patient would have decided if capable of doing so is very difficult, as it is almost impossible to completely separate one's own views and values from this decision (Brauer 2011:margin no./Rz. 12). Even when a proxy, such as a family member, knows the patient very well and has spoken with him or her about end-of-life care, the proxy's decision often does not correspond to what the patient would decide for him- or herself. Therefore, one has to bear in mind that presumed wishes are no more than a hypothetical construct and remain what the term itself suggests—an assumption. Although from a legal viewpoint a decision taken by a proxy is equivalent to one taken by the patient him- or herself and renders a medical intervention legitimate, it remains only a substitute for self-determination, even if the proxy is bound to decide according to the patient's presumed wishes.

Additionally, physicians often raise concerns about relatives being burdened with complex treatment decisions, as their emotional involvement with the patient can make it very difficult for them to act as a representative (Jox et al. 2008:62).

To meet these concerns, it is crucial that patients should be given the option of deciding whether they wish to draw up a living will containing very specific instructions or to appoint someone they trust as a healthcare proxy. Thus, self-determination trumps informal representation by family members. Moreover, procedural safeguards are needed to adequately protect incompetent patients from proxies who do not serve their interests (Jox et al. 2008:62). To this end, the decision-making power of family members has to be limited by law, and there has to be an effective public authority in charge of supervision and resolving conflicts (Explanatory Memorandum Recommendation CM/REC 2009(11), n°54).

9.4 Swiss Law

Switzerland has recently completely revised its adult protection law. The new provisions have been incorporated into the Swiss Civil Code and came into effect on 1st January 2013. The main objectives of the new adult protection law are to promote self-determination through personal provision for the future, to strengthen solidarity within the family while at the same time easing the burden on public authorities, and to provide proportionate public measures aligned with individual needs (Schweizerischer Bundesrat 2006:7011–7017).

Following the principle of self-determination, the adult protection law establishes two forms of advance directives: living wills and continuing powers of attorney. Living wills (Swiss Civil Code Articles 370–373) have to be established in writing by persons who have capacity and must be signed and dated. Minors are also allowed to establish a living will insofar as they have capacity—which means they are required to have the ability to understand what they are doing, form a free opinion and act in accordance with this opinion. According to Article 16 of the Swiss Civil Code, capacity (*Urteilsfähigkeit*) requires the *ability* to act rationally. This, however, must not be confused with a requirement that the decision itself has to be rational, which is not the case. It is possible to establish a continuing power of attorney in a living will. The person designated can be given precise instructions or more general guidelines regarding surrogate decision-making. In principle, a living will is legally binding. It can be revoked by a competent patient at any time, either verbally or in writing, or by physically destroying the document. However, when the person becomes incompetent, the physician generally has to respect the living will. The living will is not legally binding when it contravenes statutory provisions (e.g. when it requests euthanasia) or when there is reasonable doubt as to whether it expresses the patient's free will or still corresponds to the patient's presumed wishes. The latter criterion calls for the help of family members and close persons, even if the patient has established a living will. They have to be asked if there are reasonable doubts as to whether the living will still corresponds to the patient's presumed wishes, especially if the living will is ambiguous. Interpretation of living wills is crucial and has to be made with due diligence. Further, reasonable doubts may arise as to whether the living will still corresponds to the presumed wishes if the document is very old, if the patient has made conflicting remarks to family members or physicians, or if there has been significant progress in medicine since the document was written. In any case, failure to comply with the living will and the exact reasons for this decision have to be recorded in the patient's documentation. Moreover, any person close to the patient, including medical staff, can appeal to adult protection authorities and ask for a decision when a living will is not respected (Swiss Civil Code Article 373, paragraph 1, item 1).

In cases where a person has not designated a proxy him- or herself and has not established a living will, the new adult protection law specifies an order of precedence for legal proxies, according to the principle of strengthening family solidarity. The main rationale for this is that the legislative authority has determined that, in practice, family members and persons close to an incompetent adult usually do not contact adult protection services but decide on the patient's behalf without having due power of attorney. In the view of the legislative authority, it would be illusory to assume that family members would be willing to cooperate more closely with adult protection services in the future. Thus, the new adult protection law allows for the desire of family members and close persons to decide on behalf of their loved ones to be taken into account, as far as this is acceptable from an objective viewpoint and legalizes hitherto informal proxy relationships (Schweizerischer Bundesrat 2006:7013; Reusser 2012:N 47).

The order of precedence for proxies is specified in Article 378 of the Swiss Civil Code. Because the strengthening of self-determination was an important goal of the revision of the adult protection law, a proxy designated by the person concerned always takes precedence over representatives designated by law. In second place comes the legal guardian, if he or she has power of attorney in medical decisions, which has to be specifically ordered by the adult protection authority. In the absence of a designated proxy or a legal guardian, the spouse or registered partner (for same-sex couples) is authorized to take decisions on behalf of the incompetent patient. Next in line is a person who lives together with the patient; this can be a partner, a sister or a good friend. At the end of the list come children, parents and siblings.

However, kinship, legal bonds or living with the patient are not a sufficient mark of emotional closeness, and statutory orders of precedence therefore have their drawbacks, as we have already seen in Sect. 9.3. Accordingly, the Swiss law adds an important condition for legal proxies: they are required to have personally supported the patient on a regular basis. Thus, the actual existence of an intact relationship is more important than legal bonds or kinship. While this is certainly an important specification, it complicates the task of physicians: not only do they have to determine which of the family members is entitled to decide on behalf of the incompetent patient according to the statutory order of precedence, they also have to determine whether the relationship between the patient and the family member is as close as is required by law. Therefore, if it is doubtful who is entitled to consent to or refuse treatment, the adult protection authority can be asked to decide on the matter (Swiss Civil Code Article 381, paragraph 2, item 1). Moreover, if several persons (e.g. three sons) have power of attorney, the physician is entitled to assume that each acts with the agreement of the others (Article 378, paragraph 2).

Informal representatives' decisions have to be guided by the presumed wishes and the interests of the incompetent person (Article 378, paragraph 3). This provision was criticized by the Swiss National Advisory Commission on Biomedical Ethics, as it could lead to conflicts: presumed wishes, as we have seen, do not have to be reasonable from an objective viewpoint, whereas interests are usually understood in an objective sense. Therefore, the Advisory Commission recommends that the patient's interests should only be invoked if the presumed wishes cannot be determined (Brauer 2011:marginal note/Rz 21; Eichenberger and Kohler 2012:N 13). However, in a legal context, the criterion of the patient's interests is not to be understood in a strictly medical sense. In spite of being determined according to objective parameters, it includes not only medical, but also social, cultural and psychological aspects (for a similar discussion concerning the term "best interests of the child" see Michel 2009:150–152). Moreover, the criterion is important in answering the question of which medical interventions are admissible. Medical interventions that are not in the patient's interests are not admissible, even with the consent of the legal representative (Eichenberger and Kohler 2012:N 13). Thus, this criterion limits the power of the representative to give consent. On the other hand, the legal representative is allowed to refuse consent to treatments that would be in the patient's interest if this decision can be based on the patient's presumed wishes—which of course have to be established with adequate certainty when life-saving treatment is refused. In addition, the power of attorney is also limited by special statutes, e.g. the Transplantation Act or the Human Research Act.

Furthermore, the incompetent patient has to be involved in decision-making as far as this is possible (Swiss Civil Code Article 377, paragraph 3). The participation rights of the incompetent patient are based on his or her right to human dignity. The duty to involve the incompetent patient in decision-making obliges the surrogate decision-maker to consider the incompetent patient's wishes and preferences. However, only competent patients may and must bear the consequences of their decisions; therefore, if the incompetent patient's wishes and preferences cannot be reconciled with his or her interests or with what has been specified in an advance directive, his or her interests or instructions given in an advance directive will prevail.

If a healthcare proxy designated by the patient or by law does not act in the patient's interests, the physician, medical staff or any person close to the patient can appeal to the adult protection authority. This authority is also entitled to act on its own initiative.

In an emergency, there is no time to analyse living wills or contact representatives; rather, physicians have to take the necessary measures to protect the life and health of the patient. However, they also have to align their decisions with the presumed wishes and the interests of the incompetent patient, as far as these can be determined (Swiss Civil Code Article 379).

Whether or not a patient has capacity is for the physician to decide. This can be challenging, as capacity is a legal term and therefore the legal requirements for capacity have to be met. It is important to note that medical diagnosis does not determine whether or not a patient is incompetent; persons can only be declared incompetent if, at a given point in time, they lack the capacity to understand the concrete situation, form their own opinion and act according to it. Swiss law does not allow for gradation of capacity—a person either has capacity in its entirety or completely lacks it. In a decision concerning Russia (ECHR, *Shtukaturov v. Russia*, Application no. 44009/05, 27 March 2008), the ECHR criticized this approach with reference to Recommendation R(99)4 as a "common European standard" and stated that capacity should be assessed more individually. However, capacity is relative to time and subject matter: this allows for shifting capacity to be taken into account— e.g. when a patient's judgment is sometimes clear and sometimes clouded. Also, a patient may find complex decisions too challenging while still being quite capable of understanding more basic problems.

Special provision is made in the Swiss law for patients with mental disorders. Due to the increased vulnerability of these patients, proxies cannot consent to admission of an incompetent patient to a psychiatric hospital. If a mentally ill patient lacks the capacity to consent to admission him- or herself, there has to be an official commitment, or "protective placement" (*fürsorgerische Unterbringung*). The patient (and if possible a person close to the patient) has to be duly informed in writing. The patient, or a person close to the patient, can appeal against the commitment, and it has to be reviewed on a regular basis. Treatment without consent or against the expressed wishes of the incompetent patient is only legal if it is authorized by the head of department because the patient would otherwise suffer serious harm to his or her health or pose a threat to the life or physical integrity of other people, and no less invasive measure is available (Swiss Civil Code Article 434). Moreover, the patient and the trusted person have to be informed in writing about the intended treatment and can appeal against it. However, the self-determination of committed patients is limited: although advance directives do have to be taken into account, they do not have to be obeyed at all costs (Swiss Civil Code Article 433, paragraph 3). However, deviations from the living will should only be made if this is absolutely necessary because the living will cannot be reconciled with the patient's interests (Geiser and Etzensberger 2012:N 16).

9.5 Conclusions

The analysis of the relevant legal basis has shown that the law does indeed grant power of attorney to family members and close persons. This is already required by international law—as our survey of the Council of Europe's conventions, resolutions and recommendations in this field has shown—and is further elaborated by national law. In the majority of cases, this is certainly in the interests of the incompetent patient. However, certain risks and drawbacks associated with this approach have to be addressed by law.

Firstly, the decision to declare a patient incompetent is absolutely crucial. The law neither adequately specifies a procedure that has to be followed in taking this decision nor does it enumerate specific abilities, and competence is defined in quite a general manner. However, it is doubtful whether more precise legal requirements or a formal procedure for assessment of capacity would be to the best advantage of patients, as it would probably impede an appropriate response to shifting competence (relative to time and subject matter) which allows for maximum possible self-determination. The responsibility to assess this competence lies with physicians and is certainly a major one. Nevertheless, ethical guidelines and best practice standards seem to be a more suitable approach for meeting this responsibility than the development of a rather inflexible legal definition.

Secondly, self-determination should prevail. Therefore, the designation of a proxy by the patient him- or herself or the execution of a living will should be more important than representation provided by law.

Thirdly, national law should regulate who is entitled to take decisions on behalf of an incompetent patient in the absence of a designated proxy, as this cannot be the physician conducting the treatment, except for emergency cases. The power of the representative has to be limited by law, and there have to be legal guidelines regulating the decision-making process and the duty to involve the incompetent patient. Moreover, legal safeguards have to be provided in case the representative or medical staff do not act in the patient's interests or according to his or her presumed wishes. On the one hand, this can be monitored by the medical staff in direct contact with the representative and, on the other hand, by the legal representative. However, as the medical staff should not be entitled to overrule the representative's decision except in emergency cases, this requires the existence of an efficient and professional public authority, responsible for making a decision in cases of conflict. This decision must be open to appeal and, therefore, the legal process has to be open to the persons concerned. Thus, the interests of incompetent patients can be protected by the persons close to them and by their caregivers.

References

Andorno, Roberto. 2010. Editorial. An important step in the promotion of patients' self-determination. *European Journal of Health Law* 17: 119–124.

Andorno, Roberto. 2011. Regulating advance directives at the Council of Europe. In *Self-determination, dignity and end-of-life care. Regulating advance directives in international and comparative perspective*, ed. Stefania Negri, 73–86. Leiden/Boston: Martinus Nijhoff Publishers.

Brauer, Susanne. 2008. Die Autonomiekonzeption in Patientenverfügungen – Die Rolle von Persönlichkeit und sozialen Beziehungen. *Ethik in der Medizin* 20(3): 230–239.

Brauer, Susanne. 2011. Patientenverfügung und Demenz im neuen Erwachsenenschutzrecht aus Sicht der Ethik. *Jusletter* vom 29 August 2011.

Council of Europe. 1950. Convention for the protection of human rights and fundamental freedoms. http://conventions.coe.int/treaty/en/treaties/html/005.htm

Council of Europe. 1997a. Convention for the protection of human rights and dignity of the human being with regard to the application of biology and medicine: Convention on human rights and biomedicine. http://conventions.coe.int/Treaty/en/Treaties/Html/164.htm

Council of Europe. 1997b. Explanatory report. http://conventions.coe.int/Treaty/EN/Reports/Html/164.htm

Council of Europe. 1999a. Recommendation no. R(99)4. http://www.coe.int/t/dg3/healthbioethic/texts_and_documents/Rec(99)4E.pdf

Council of Europe. 1999b. Recommendation no. 1418 (1999). Protection of the human rights and dignity of the terminally ill and the dying. http://assembly.coe.int//main.asp?link=http://assembly.coe.int/Documents/AdoptedText/ta99/erec1418.htm#1

Council of Europe. 2009. Recommendation CM/Rec(2009)11 of the Committee of Ministers to the member states on principles concerning continuing powers of attorney and advance directives for incapacity. https://wcd.coe.int/ViewDoc.jsp?id=1563397&Site=CM

Council of Europe. 2012. Parliamentary Assembly Resolution 1859. http://assembly.coe.int/Main.asp?link=/Documents/AdoptedText/ta12/ERES1859.htm

Eichenberger, Thomas, and Theres Kohler. 2012. Kommentierung von Art. 378 ZGB. In *Basler Kommentar Erwachsenenschutz*, ed. Thomas Geiser and Ruth Reusser. Basel: Helbing Lichtenhahn.

Frowein, Jochen A., and Wolfgang Peukert. 2009. *Europäische Menschenrechtskonvention. EMRK-Kommentar*. 3. Auflage. Berlin: Engel Verlag, Kehl.

Geiser, Thomas, and Mario Etzensberger. 2012. Kommentierung von Art. 433 ZGB. In *Basler Kommentar Erwachsenenschutz*, ed. Thomas Geiser and Ruth Reusser. Basel: Helbing Lichtenhahn.

Gevers, Sief, Joseph Dute, and Herman Nys. 2012. Surrogate decision-making for incompetent elderly patients: The role of informal representatives. *European Journal of Health Law* 19: 61–68.

Grabenwarter, Christoph, and Katharina Pabel. 2012. *Europäische Menschenrechtskonvention*. 5. Auflage. Basel: Helbing Lichtenhahn.

Herdegen, Matthias. 2012. *Völkerrecht*. 11. Auflage. München: C.H. Beck.

Jox, Ralf J., Sabine Michalowski, Jorn Lorenz, and Jan Schildmann. 2008. Substitute decision making in medicine: Comparative analysis of the ethico-legal discourse in England and Germany. *Medicine, Health Care & Philosophy* 11: 153–163.

Michel, Margot. 2009. *Rechte von Kindern in medizinischen Heilbehandlungen*. Basel: Helbing Lichtenhahn.

Reusser, Ruth E. 2012. Allgemeine Vorbemerkungen. In *Basler Kommentar Erwachsenenschutz*, ed. Thomas Geiser and Ruth Reusser. Basel: Helbing Lichtenhahn.

Schweizerischer Bundesrat. 2006. Botschaft zur Änderung des Schweizerischen Zivilgesetzbuches (Erwachsenenschutz, Personenrecht und Kindesrecht) vom 28. Juni 2006 (BBl 2006 7001). http://www.admin.ch/ch/d/ff/2006/7001.pdf

Zellweger, Caroline, Susanne Brauer, Christopher Geth, and Nikola Biller-Andorno. 2008. Patientenverfügungen als Ausdruck individualistischer Selbstbestimmung? Die Rolle der Angehörigen in Patientenverfügungsformularen. *Ethik in der Medizin* 20(3): 201–212.

Chapter 10
Advance Directives and the Ethos of Good Nursing Care

Settimio Monteverde

10.1 Nurses and Advance Directives: Framing the Issue

In post-industrial countries with an extensive health and service sector, nurses represent the largest part of the healthcare workforce (McHugh et al. 2011). Advance directives have emerged as a topic of theoretical and practical concern in the context of nursing practice as well as for other healthcare professions (Wareham et al. 2005). Since the late 1970s, these documents have been increasingly recognized by the law, professional bodies and healthcare ethicists as valid means of exercising self-determination in situations of incompetence (Silveira et al. 2010). Even so, completion rates vary considerably. They are influenced by geographical, social, economic, racial, ethnic and literacy factors (see, for the US, Winter et al. 2010; Rosnick and Reynolds 2003; Gerst and Burr 2008; Kwak and Haley 2005; for Germany, Evans et al. 2012) and also by health status (Sahm et al. 2005). Nonetheless, thanks to the growing ethico-legal consensus and clinical acceptance, advance directives are accorded authoritative force in directing surrogate decision-making (Andorno et al. 2009).

Basically, the possibility of stating treatment preferences in advance for situations of incompetence rests on the assumption that respect for autonomy also encompasses respect for *precedent* autonomy (Davis 2002; Blondeau et al. 2000). This is also undisputed in nursing ethics, as is evident, for example, from the Code of Ethics of the American Nurses' Association (2001):

> Nurses actively participate in assessing and assuring the responsible and appropriate use of interventions in order to minimize unwarranted or unwanted treatment and patient suffering. The acceptability and importance of carefully considered decisions regarding resuscitation status, withholding and withdrawing life-sustaining therapies, forgoing

S. Monteverde (✉)
Institute of Biomedical Ethics, University of Zurich, Zurich, Switzerland
e-mail: settimio.monteverde@uzh.ch

P. Lack et al. (eds.), *Advance Directives*, International Library of Ethics, Law, and the New Medicine 54, DOI 10.1007/978-94-007-7377-6_10, © Springer Science+Business Media Dordrecht 2014

medically provided nutrition and hydration, aggressive pain and symptom management and advance directives are increasingly evident. The nurse should provide interventions to relieve pain and other symptoms in the dying patient even when those interventions entail risks of hastening death.

10.2   The Impact of Advance Directives

10.2.1   The Bigger Picture

Various ethical, legal and clinical aspects of advance directives are subject to ongoing debate within biomedical ethics. One example is the question of whether or how precedent autonomy is compatible with existing standards for the validity of informed consent in other areas of medical and nursing practice, such as surgery, anaesthesia, self-discharge of competent patients or referrals to specialized clinics. Another is the predictability of treatment options and preferences for dementia care (see Davis 2004; Muramoto 2011). But in spite of these unresolved issues (also discussed by nursing scholars), advance directives have had a profound impact on the day-to-day clinical work of all healthcare professionals, including nurses (Adams et al. 2011). This is mainly due to the following factors:

- the broad availability and accessibility of written forms and counselling services enabling individuals to complete advance directives
- legal regulations governing the binding character of advance directives in an increasing number of countries
- the social and ethical acceptance of precedent autonomy being exercised in a variety of situations where decision-making capacity is lost (from imminent death to dementia and from acute to chronic care settings).

In order to understand the impact of advance directives on healthcare systems and on the individuals, professions and institutions within them, it is essential to focus not just on patients, but on a wide range of outcomes relating to the health and well-being of individuals. Of course, in implementing advance directives, compliance with the patient's prior wishes and with individualized care plans is undoubtedly one of the most critical outcome variables, insofar as it relates to the ultimate aim of advance directives themselves. But a closer look at the literature reveals a more complex picture, with data that are far from consistent. Nevertheless, these data are helpful in understanding and explaining the perspective of nurses.

Older observational studies (e.g. Connors et al. 1995; Prendergast 2001) and interventional studies (Thompson et al. 2003) present quite a sobering picture of the impact of advance directives on patient care, with low correlations between the patient's previously expressed wishes and the care actually delivered, and a restrained attitude among professionals towards executing patients' known preferences. However, a large US multisite study—while not specifically addressing advance directives—revealed the importance of communication: end-of-life discussions between physicians, patients and caregivers resulted in fewer aggressive interventions in the final week of life (Wright et al. 2008). Other, more recent studies involving educational interventions for patients, proxies and/or professionals show better outcomes at various levels, e.g. knowledge of patients' wishes, attitudes of professionals towards advance directives, and numbers of advance directives completed and/or executed (Detering et al. 2010; Silveira et al. 2010; Schickedanz et al. 2009; in der Schmitten et al. 2011; see also Chap. 13). The best outcomes in terms of compliance with previously stated treatment preferences are described in studies using educational interventions in which professionals depart from paradigmatic hypothetical scenarios. These studies suggest that—apart from ethical and legal issues and factors relating to individual preferences—*environmental conditions* of daily practice in healthcare institutions exert a powerful influence on advance care planning, and therefore also on the implementation of advance directives (Thompson et al. 2003). This is confirmed by the nursing literature (e.g. Georges and Grypdonck 2002; Pavlish et al. 2011; Wolf and Zuzelo 2006).

10.2.2 Building Up the Research

Investigations of the roles and attitudes of nurses involved in advance care planning and the implementation of advance directives can be divided into three different— though overlapping—phases.

In a general sense, the work of nurses caring for patients at the end of life has only been subject to scholarly investigation in the last three decades (for reviews see Adams et al. 2011; Benbenishty et al. 2006; Georges and Grypdonck 2002; Ryan and Jezewski 2012). This research sought not only to gather evidence about the multifaceted responsibilities of nurses caring for the severely ill and the dying, but also to shed light on their roles and role conflicts within the multiprofessional healthcare team. In a subsequent phase, following the introduction of new regulations on advance directives in the US, Canada and Europe, many scholars investigated the impact of these new legal provisions on nurses in specific countries or states (Blondeau et al. 2000; Silveira et al. 2010; Wareham et al. 2005; Barrio-Cantalejo et al. 2009; Walerius et al. 2009). Finally, in a third phase, the impact of advance directives on specific nursing and patient populations has been assessed by focusing on nursing specialties such as home health, hospice, intensive, dialysis and emergency care, or clinical nurse specialists (Badzek et al. 2006; Scherer et al. 2006; Hamric and Blackhall 2007; Meehan 2009). All three phases have contributed to our understanding of the bigger picture.

10.3 A Systemic Perspective

Many nursing scholars have not confined their analyses to the range of values and attitudes held by *individual* nurses towards advance directives, or to the question of how these values and beliefs may be modified by particular variables (e.g. adherence to advance directives in terminal stages of cancer versus adherence in moderate Alzheimer's disease). Certainly this ethical dimension is of great importance insofar as it brings to light central elements of the ethos of nursing— how care dependency, precedent autonomy, quality of life, physical and existential suffering and surrogacy are viewed, and what rights and duties can be ascribed to professionals, patients, proxies and relatives. Rather than limiting their obser-vations to individual ethics and personal attitudinal factors, some authors advocate a systemic perspective and adopt an organizational ethics viewpoint (e.g. Ryan and Jezewski 2012; Adams et al. 2011; Hamric and Blackhall 2007; Johns 1996; see also Brett 2002). When the issue of advance directives is addressed from this viewpoint, the main focus is placed on the institutional conditions under which nurses, physicians and co-workers act, advance directives are discussed and executed, and healthcare is provided for incompetent patients.

From a nursing point of view, taking a systemic perspective on advance directives basically means understanding the whole process of advance care planning as a systemic effort involving all the professions in charge of the patient. It comprises a series of steps:

1. *Emphasizing the importance of communication*: Patterns of communication between different healthcare professions, providers, patients and proxies in the various stages of the advance care planning process have to be carefully analysed (Barrio-Cantalejo et al. 2009).
2. *Addressing issues of power*: Imbalances of power within the healthcare team have to be recognized as potential sources of tension between physicians responsible for decision-making and nurses responsible for implementation. Evidence shows that, when patients' treatment preferences are not followed, standards of care are not met and nurses are not in a position to change the course of action adopted, the resulting tensions can lead to considerable moral distress for nurses unless these situations are explicitly addressed and critically discussed (Hamric and Blackhall 2007; Peter and Liaschenko 2004).
3. *Creating opportunities for collaborative learning*: Continuous professional development activities such as "death rounds" or regular case discussions on the ward have to be taken as opportunities to address these issues (Puntillo and McAdam 2006). The aim of these activities is to increase mutual knowledge of other professions in charge of the same patients, to overcome stereotypes that act as barriers to collaboration (Ateah et al. 2011) and to enhance the understanding of moral diversity within the healthcare team (Irvine et al. 2002).
4. *Striving for continuous adaptation and refinement of care plans*: The continuous involvement of patients, as well as their legal representatives, in decision-making by the healthcare team (Richter and Eisemann 1999) requires ethical

sensitivity, awareness and alertness of all team members to changes in health status and to statements made by incompetent patients as well as opinions expressed by parents and proxies.

As regards the fourth point, nurses are very well acquainted with methods of adapting nursing care to changing circumstances. These are described in a paradigmatic way in the so-called *nursing process*, a widespread methodological approach in educational and clinical contexts that fosters critical and adaptive thinking (Roper et al. 2000; Hackmann 2011). This circular problem-solving method essentially involves the following steps: *assessing* and *diagnosing* the problem, *identifying* meaningful outcomes, *planning* a course of action (as a rule a nursing intervention), *implementing* a plan and *evaluating* the outcomes of the intervention. The utility of the nursing process has also been increasingly recognized for ethical issues (Tschudin 2002; see also Turkel et al. 2012). It can thus be applied in adapting advance care plans to new circumstances—foreseeable or not. As part of the wider assessment, it is also necessary to identify ethical aspects relating to precedent autonomy, conflicting moral intuitions among different (professional and lay) stakeholders and other ethical principles. In sum, the systemic perspective proposed in the nursing literature expands the range of outcomes of advance directives from a narrow patient orientation to further variables. These include communication, collaboration and education, as well as the identification of enabling and hindering factors from the perspectives of providers, professionals, patients, their families and proxies. A systemic perspective locates advance directives in a more extensive web of patient-centred, sustainable and responsible decision-making (Malloy et al. 2009).

10.4 Conditions in the Field

10.4.1 Between Authority and Autonomy

Initial research showed that, in relation to advance care planning and end-of-life care, the work environment of nurses was essentially shaped by four factors:

1. The use of technological and pharmaceutical advances to sustain life was regarded as a therapeutic imperative even in conditions deemed medically futile (Wolf and Zuzelo 2006; Malone 2003).
2. Legal uncertainties about the liability of doctors who withhold or withdraw life-sustaining treatment resulted in a perceived risk of overtreatment by physicians wishing to be "on the safe side" (Johnstone 2009).
3. The ethical question of when a human being should be allowed to die and who should ultimately be responsible for answering this question had yet to be settled by medical and nursing ethicists and by professional bodies, leading to considerable moral uncertainty in clinical practice.

4. Nurses complained of poor communication on these issues within the healthcare team—especially with the physicians in charge—and a perceived mismatch between their limited participation in medical decision-making and their extensive responsibilities in carrying out medical instructions (Oberle and Hughes 2001; Brett 2002).

These four factors have been reported to contribute to the moral distress of nurses caring for patients with chronic conditions and at the end of life. They are of particular relevance in dealing with advance directives in the process of planning and delivering care. They are discussed as potential—though not the only—sources of moral concern.

While the qualitative and quantitative data available paint a consistent picture and metrics have been developed to assess ethical climate, moral distress and collaborative decision-making (Corley et al. 2001; Pauly et al. 2009; Hamric et al. 2012), the reasons proposed to explain these phenomena vary widely. This makes it difficult to draw clear inferences from observed outcomes (e.g. moral distress, failure to meet adequate standards of care) to predisposing factors. A methodological challenge is posed by the predominance of small-scale (one-time, one-site, single-cohort) descriptive studies due to limited generalizability (Hamric and Blackhall 2007).

10.4.2   Morally Unambiguous Versus Ambiguous Situations

If the field conditions described above are considered in the context of developments in ethics, law and medical reasoning, a tentative distinction can be made between two different scenarios describing possible "relationships" between nurses and advance directives.

The first type of scenario is *morally unambiguous*, in that nurses are confronted with a situation which they cannot change and which *clearly* offends their moral sensibilities. Examples are situations where advance directives are explicitly disregarded or ignored (for whatever reasons), standards of care are not met, patients are suffering existentially, or nurses' requests for help are not answered. In a study of moral stress and burnout, Severinsson (2003:61) reports:

> The continuity in the relationship with the patient and family members, and the predictability of death, can generate feelings of inadequacy. As a result, the professional simply stops identifying ethical problems because one can no longer cope with all the ethical problems encountered. The need to discuss ethical problems with colleagues and to obtain ethical advice was expressed. Without doubt, there is a need for a response from others, whose scope for action is extremely limited. This was expressed by the nurse as lack of support, as well as a desire for confirmation and mutuality in the relationship with the supervisor and other colleagues.

By contrast, the second type of situation nurses are confronted with does not involve legal uncertainty, an imbalance of power or poor communication, but a *real* ambiguity concerning the content and binding force of advance directives (e.g. a "radical" refusal of life-preserving measures such as oral antibiotics in moderate Alzheimer's disease). These situations can be described as *morally ambiguous* insofar as they are an expression not of an abuse or misuse of power, but of "true" ethical complexity (Hood 2012) which lies in the nature of the circumstances. In this connection, Mezey et al. (1996: 09) comment:

> It is important to know the facility's method for resolving conflicts between family members and the patient, the patient/family and care providers, or health care professionals. If a patient or proxy's treatment choice conflicts with a nurse's beliefs and the nurse cannot offer care in accordance with those wishes, the appropriate person within the facility should be notified. The overriding concern should be that the patient's care wishes be followed. Patients should be cared for by a nurse who can follow their wishes or the patient should be transferred to another facility.

It has been shown that it is of great importance for nurses and co-workers to distinguish between morally unambiguous and ambiguous situations when implementing advance directives. While the former involve a high risk of moral distress and call for measures to prevent burnout and moral disengagement, the latter may be resolved by adopting elements of the systemic approach described above.

10.5 Dimensions of Nursing Care

In order to fully understand the significance of advance directives for the nursing profession, it is necessary to consider the wider framework of nurses' contributions to end-of-life care and dementia care—the two main clinical contexts in which advance directives are implemented (Silveira et al. 2010; Maust et al. 2008; Triplett et al. 2008). Many authors describe the pivotal role played by nurses in the advance planning and delivery of care for patients who have lost decision-making capacity (Adams et al. 2011; Benbenishty et al. 2006). Although not exclusively for end-of-life care, the space in which nurses act can be seen as constituted by three axes:

- *closeness*: the physical work environment of nurses and their proximity to the patient, the provision of basic care and treatment care, and their close contact with relatives, surrogates, significant others and members of the healthcare team (Nortvedt 2001)
- *holism*: the professional understanding of nursing as a holistic activity, integrating physical, psychosocial and spiritual aspects of birth, growth, falling ill, recovering, suffering and dying (Gustafsson et al. 2007) and considering nursing care as a specific response to different needs of human beings across the lifespan
- *ethos*: moral understandings of nursing work, referring to traditions that value relationships, vulnerabilities of agents, and mutual rights and duties arising from them, thus seeing nursing as an essentially moral enterprise (Turkel et al. 2012; Sellman 2011)

Closeness, holism and the moral value of relationships—these three distinctive features certainly explain the substantial contribution of nurses to the advance planning and delivery of patient-centred care, as well as their roles within multiprofessional healthcare teams. This also applies to the specific context of preparing, updating and implementing advance directives (Meehan 2009). However, the privileged position of nurses deriving from closeness, holism and the ethos of good nursing care is not free of ambiguity in the context of advance care planning by means of advance directives.

10.6   Flexible Workers in (Too) Open Spaces

In general terms, Peter and Liaschenko (2004:222) describe the challenges associated with proximity and the striving for holism (in the sense of knowing the patient's particular needs) as a fundamental problem inherent in nursing ethics:

> Without proximity the knowledge of the particular needs of the other that an ethic of care demands would probably not be possible. Defining nursing and nursing ethics in terms of the proximity of nurses to patients, however, can lead to ambiguity. Human needs can often be unpredictable and endless. Again, as Bauman (1993) has claimed, in a state of proximity responsibility is "unlimited". Being the one there, or the one close enough to provide care, brings with it the moral complexity of being responsible for the open-ended nature of human needs. Human care-giving, such as that which takes place in organized health-care, requires that at least some care-givers be flexible in their roles; if not, many needs of patients would not be met. Traditionally, the flexible worker has been [. . .] the nurse.

For these authors, "flexibility" is an ambivalent concept, and it remains unclear whether it should be considered a virtue or a vice—i.e. an expression of a strong or a weak character. However, it is considered to be essential. First of all, when nurses deal with advance directives at different stages of the advance care planning process, flexibility can be seen as an "external" requirement: the working environment requires nurses to perform multiple tasks within the healthcare institution—such as housekeeping, generic nursing care, treatment care, counselling of patients and relatives, teaching students, etc.—and to continuously adjust or reconcile these tasks with the patients' needs. These needs are known to the nurse as a result of her/his proximity to the patient. This is the "suspicious" component of flexibility, opening up potentially unlimited responsibility of the nurse for the individual patient.

More significantly, this does not happen by chance, but because of the way the system is designed: not only do superiors or other professionals expect nurses to conform to these requirements, but also patients and, last but not least, the nurses themselves. The external flexibility resulting from the closeness of the patient-nurse interaction poses a challenge for the nurse—to advocate or even defend the patient's declared or perceived needs within the system and vis-à-vis its stakeholders. The "unsuspicious" component of this kind of flexibility is that the underlying attitude is perfectly compatible with the moral understanding of nursing as a patient-oriented profession (e.g. American Nurses' Association 2001).

A large part of the literature on advance directives addresses nurses' ethical dilemmas arising from this kind of flexibility: qualitative and quantitative data indicate that failure to honour patients' previously declared wishes results not only in over- or undertreatment, but also in significant moral distress for nurses, who often see themselves not merely as witnesses but as perpetrators or accessories to decisions they cannot support (Brett 2002; Hamric and Blackhall 2007; Johnstone 2009; McHugh et al. 2011; Oberle and Hughes 2001). External flexibility considers the nurse as the patient's companion, compensating for the unpredictabilities of complex healthcare systems by ascertaining the patient's needs through *relational knowing* (Wright and Brajtman 2011, see also Moser et al. 2010). In sum, external flexibility locates the nurse and the patient in a relationship of proximity comprising both spatiotemporal and moral aspects, in the sense that the nurse's and the patient's values regarding precedent autonomy *converge* (see Peter and Liaschenko 2004). Here, the nurse clearly assumes the role of the patient's advocate. The impossibility of acting in accordance with the patient's previously stated wishes is therefore perceived as a source of moral distress.

Interestingly, in discussing advance directives and nursing, many scholars report that nurses establish a semantic connection between the two expressions not in terms of autonomy, but of advocacy (e.g. Ryan and Jezewski 2012; Alfonso 2009). Nevertheless, some authors (e.g. Blondeau et al. 2000:407) view the concepts of autonomy and advocacy not as mutually exclusive, but as complementary:

> This conviction that the patient's autonomy should be respected brings with it, however, a particular requirement: that of exercising the role of advocacy. This role can take different forms: informing patients of their rights; ensuring that patients have all the information necessary to make enlightened choices; supporting patients in their decisions; and protecting patients' interests. We should stress that exercising a role of advocacy represents an ethical obligation. In the context of advance directives, the results of this study indicate that certain lapses can compromise the adequate exercise of the role of advocacy.

In addition to the external form of flexibility, an internal form can be identified: internal flexibility requires nurses to realize that the caring relationship does not imply a convergence of values between nurse and patient, but that this relationship can also persist within value diversity. Here, what has to be recognized and accepted is the tension between physical closeness and "distance" in values, or between "proximal" and "distal" nursing (Peter and Liaschenko 2004; Malone 2003).

In delivering care for people who lack decision-making capacity, nurses can be confronted with situations in which not *failing to honour*, but *executing* patients' preferences can diverge from their own moral values and beliefs (e.g. with regard to dementia care and withdrawal of fluids or antibiotics). In such situations, it is vital to cope with these tensions and to guarantee a continuity of care (Ryan and Jezewski 2012; Mezey et al. 1996). The root causes of such tensions are to be seen not only in diverging personal values but also in the different traditions shaping the *ethos of nursing* itself:

- These traditions reflect day-to-day experiences of caring for demented, mentally disabled or brain-injured patients. Specifically, they address the epistemic problem of the value of the "lived body" and of bodily experiences, as well as the significance of bodily expressions (see Huber 2006).
- They provide a language which considers phenomenology as offering valuable access to the lifeworld of patients who have lost decision-making capacity (Benner 1994).
- They highlight the interactions between the principles of respecting autonomy and preserving bodily integrity and criticize tendencies towards disembodiment and "rationalistic" reduction in contemporary bioethics (Leach Scully 2008; O'Neill 2002).
- Finally, they propose alternative models of "embodied", "relational", "narrative", "culturally adapted" or "situated" autonomy (see Wareham et al. 2005; Schwerdt 1998).

Knowledge of the origins of the problems experienced by nurses with advance directives can help to avoid these "perils of proximity" (Peter and Liaschenko 2004). It can help to recognize situations of moral distress in which the basic conflict does not arise from differing moral views between individuals (e.g. nurse and patient at the time of setting up an advance directive), but from *rival conceptions of autonomy within the ethos of nursing itself*. Of course, this knowledge cannot "resolve" the tension of moral distress, but it may mitigate it by offering a plausible explanation.

10.7 Advance Directives: Clarifying Nurses' Roles

10.7.1 Allocating the Tasks

As stressed by many authors, advance care planning involves a series of steps and processes that have to be carefully prepared and executed in order to reach the desired goals and outcomes at the individual and systemic level (Detering et al. 2010; in der Schmitten et al. 2011). It is essential to clarify the role of nurses in the context of advance directives, given the complexity of clinical situations, legal provisions and institutional guidelines, as well as patient preferences, values, therapeutic measures and possible courses of illnesses, which may or may not be predictable.

Nurses can adopt a variety of roles within the process of setting up, updating and implementing advance directives. These require counselling and listening skills and also sound clinical judgment, not only in executing the patient's wishes, but in negotiating meaningful alternatives with physicians, proxies and the legal representatives when clinical, ethical and legal evidence is given.

10.7.2   Overcoming the "Doctor-Nurse Game" Through Education

When one reviews the research presented in the previous sections and looks ahead to future developments, caution seems to be called for in one area. There is no doubt that helping patients to clarify their own values concerning issues of life, death and care dependency is an eminently interprofessional effort, as is the execution of advance directives in situations of incompetence. However, it is not easy to define the extent to which every individual nurse can or must be involved in the process of drafting advance directives. The nature of the informed consent given through advance directives remains unclear in many respects (see Davis 2004). But the literature tends to consider it in analogy to the patient-physician encounter and therefore also in analogy to some kind of *treatment contract*, with concomitant responsibilities and liabilities. The legal prerequisite for medical interventions—the informed consent given within a treatment contract between physician and patient—seems to be formally upheld also for *anticipated* treatments. This is especially evident for situations of pre-existing illness, in which advance directives contain specific care and medication plans (e.g. symptom control measures).

   Given these legal aspects, the extent to which nurses can be involved in the process of patient counselling and drafting advance directives is less a question of general rules, but more a question of education and experience, and also of careful negotiation and meaningful division of tasks between nurses, physicians and social workers. At present, clinical nurse specialists (CNS), advance practice nurses (APN) or advanced nursing practitioners (ANP) working with specific patient populations are most likely to exercise such competencies. At present, this would appear to be more feasible in the US, where such models have been developed and successfully implemented, than in Europe, where these models are now becoming more widespread. Models of meaningful divisions of tasks between physicians and specialized nurses (generally operating at a systemic level) for advance directives counselling are discussed in Meehan (2009) and Grace (2009).

   As mentioned above, specialized nurses (e.g. CNS, APN and ANP) are most likely to be qualified and to have the resources necessary for this task. But what does this mean? Generally speaking, helping patients and healthy people to drawing up advance directives is not a matter of professional entitlement, power or privilege. It is basically a matter of professional responsibility based on educational preparedness. Such preparedness covers quite complex issues of ethics, communication, projected courses of illness, legal frameworks, self-determination and the self-understanding of individuals situated in communities of shared values. A cursory look at existing nursing curricula raises doubts as to whether such preparedness can be generally presupposed, which of course also holds true for physicians. It would be fatal to misuse advance directives counselling as a playground for the infamous "doctor-nurse game" (Stein and Wis 1967; Wagner 2011). But as the field conditions described above suggest, this cannot be ruled out for the past decades, and the results are detrimental not only to the working environment, but especially to the quality of care provided to patients. Advance directives should not be handled as surrogate arenas for clarification of professional boundaries, but basically with a view to their potential impact on patients for situations of incompetence. This calls for a systemic effort in which nurses have proved to be key players for a variety of reasons, as will hopefully also be the case in the future.

10.8 Conclusions

The literature shows that nurses have unique possibilities to support patients in the whole process of clarifying their values and wishes, facilitating and documenting the expression of their preferences, and performing patient-centred care in conditions associated with incompetence, be they chronic, acute or at the end of life. However, it is important to acknowledge not only the advantages of proximity of nursing care to the lifeworld of patients, relatives and significant others, but also the associated challenges. The most important one is to withstand the tension between closeness and distance, and to resist the temptation to resolve this tension either by ignoring advance directives or by failing to question them if necessary. Withstanding the tension means striving to recognize what best respects the patient's wishes (whether this be in accordance with or against the professional's own values or beliefs) and is in the patient's best interests.

Owing to their numerical strength, their closeness to patients, families and members of the multiprofessional team, and the holistic claims of their work, nurses are in a position to act as systemic change agents in establishing and maintaining a culture of trust and patient orientation at every stage of the advance planning and delivery of care—with or without advance directives. Nurses have important tasks in the consensual planning and delivery of medical and nursing care, in evaluating these measures, and in reporting their observations, intuitions and ethical values within the multiprofessional team. Like other members of the healthcare team, they also have the right and the duty to express their ethical concerns regarding care-related conflicts in order to foster collaborative interdisciplinary learning.

References

Adams, Judith A., Donald E. Bailey, Ruth A. Anderson, and Sharron L. Docherty. 2011. Nursing roles and strategies in end-of-life decision making in acute care: A systematic review of the literature. *Nursing Research and Practice* 2011: 527834. doi:10.1155/2011/527834.

Alfonso, Heather. 2009. The importance of living wills and advance directives. *Journal of Gerontological Nursing* 35(10): 42–45.

American Nurses' Association. 2001. Code of ethics for nurses with interpretive statements. http://www.nursingworld.org/MainMenuCategories/EthicsStandards/CodeofEthicsforNurses/Code-of-Ethics.pdf. Accessed 12 July 2012.

Andorno, Roberto, Nikola Biller-Andorno, and Susanne Brauer. 2009. Advance health care directives: Towards a coordinated European policy? *European Journal of Health Law* 16(3): 207–227.

Ateah, Christine A., Wanda Snow, Pamela Wener, Laura MacDonald, Colleen Metge, Penny Davis, Moni Fricke, Sora Ludwig, and Judy Anderson. 2011. Stereotyping as a barrier to collaboration: Does interprofessional education make a difference? *Nurse Education Today* 31 (2): 208–213. doi:10.1016/j.nedt.2010.06.004.

Badzek, Laurie A., Nan Leslie, Renee U. Schwertfeger, Pamela Deiriggi, Jacqueline Glover, and Laura Friend. 2006. Advanced care planning: A study on home health nurses. *Applied Nursing Research* 19(2): 56–62. doi:10.1016/j.apnr.2005.04.004.

Barrio-Cantalejo, Inés M., Adoración Molina-Ruiz, Pablo Simón-Lorda, Carmen Cámara-Medina, Isabel Toral López, Maria del Mar Rodríguez del Aguila, and Rosa M. Bailón-Gómez. 2009. Advance directives and proxies' predictions about patients' treatment preferences. *Nursing Ethics* 16(1): 93–109. doi:10.1177/0969733008097995.

Bauman, Z. 1993. *Postmodern ethics*. Oxford: Blackwell Publishing.

Benbenishty, Julie, Freda D. Ganz, Anne Lippert, Hans-Henrik Bulow, Elisabeth Wennberg, Beverly Henderson, Mia Svantesson, et al. 2006. Nurse involvement in end-of-life decision making: The ETHICUS study. *Intensive Care Medicine* 32(1): 129–132. doi:10.1007/s00134-005-2864-1.

Benner, Patricia E. 1994. *Interpretive phenomenology: Embodiment, caring, and ethics in health and illness*. Thousand Oaks: Sage Publications.

Blondeau, Danielle, Mireille Lavoie, Pierre Valois, Edward W. Keyserlingk, Martin Hébert, and Isabelle Martineau. 2000. The attitude of Canadian nurses towards advance directives. *Nursing Ethics* 7(5): 399–411.

Brett, Allan S. 2002. Problems in caring for critically and terminally ill patients: Perspectives of physicians and nurses. *HEC Forum* 14(2): 132–147.

Connors, Alfred F., Neil V. Dawson, Norman A. Desbiens, William J. Fulkerson, Lee Goldman, William A. Knaus, Joanne Lynn, et al. 1995. A controlled trial to improve care for seriously ill hospitalized patients: The study to understand prognoses and preferences for outcomes and risks of treatments (SUPPORT). *JAMA: The Journal of the American Medical Association* 274 (20): 1591–1598. doi:10.1001/jama.1995.03530200027032.

Corley, Mary C., Ronald K. Elswick, Martha Gorman, and Theresa Clor. 2001. Development and evaluation of a moral distress scale. *Journal of Advanced Nursing* 33(2): 250–256. doi:10.1111/j.1365-2648.2001.01658.x.

Davis, John K. 2002. The concept of precedent autonomy. *Bioethics* 16(2): 114–133.

Davis, John K. 2004. Precedent autonomy and subsequent consent. *Ethical Theory and Moral Practice* 7(3): 267–291.

Detering, Karen M., Andrew D. Hancock, Michael C. Reade, and William Silvester. 2010. The impact of advance care planning on end of life care in elderly patients: Randomised controlled trial. *BMJ* 340(mar23 1): c1345. doi:10.1136/bmj.c1345.

Evans, Natalie, Claudia Bausewein, Arantza Meñaca, Erin V.W. Andrew, Irene J. Higginson, Richard Harding, Robert Pool, and Marjolein Gysels. 2012. A critical review of advance directives in Germany: Attitudes, use and healthcare professionals' compliance. *Patient Education and Counseling* 87(3): 277–288. doi:10.1016/j.pec.2011.10.004.

Georges, Jean-Jacques, and Mieke Grypdonck. 2002. Moral problems experienced by nurses when caring for terminally ill people: A literature review. *Nursing Ethics* 9(2): 155–178.

Gerst, Kerstin, and Jeffrey A. Burr. 2008. Planning for end-of-life care: Black-white differences in the completion of advance directives. *Research on Aging* 30(4): 428–449. doi:10.1177/0164027508316618.

Grace, Pamela J. 2009. *Nursing ethics and professional responsibility in advanced practice*. Boston: Jones and Bartlett Publishers.

Gustafsson, Christine, Margareta Asp, and Ingegerd Fagerberg. 2007. Reflective practice in nursing care: Embedded assumptions in qualitative studies. *International Journal of Nursing Practice* 13(3): 151–160. doi:10.1111/j.1440-172X.2007.00620.x.

Hackmann, Mathilde. 2011. Zur Geschichte des Pflegeprozesses. Justitia half dem Klassiker auf die Sprünge. *Pflege Zeitschrift* 64(9): 557–559.

Hamric, Ann B., and Leslie J. Blackhall. 2007. Nurse-physician perspectives on the care of dying patients in intensive care units: Collaboration, moral distress, and ethical climate. *Critical Care Medicine* 35(2): 422–429. doi:10.1097/01.CCM.0000254722.50608.2D.

Hamric, Ann B., Christopher T. Borchers, and Elizabeth G. Epstein. 2012. Development and testing of an instrument to measure moral distress in healthcare professionals. *AJOB Primary Research* 3(2): 1–9. doi:10.1080/21507716.2011.652337.

Hood, Rick. 2012. A critical realist model of complexity for interprofessional working. *Journal of Interprofessional Care* 26(1): 6–12. doi:10.3109/13561820.2011.598640.

Huber, Lara. 2006. Patientenautonomie als nichtidealisierte "natürliche Autonomie". *Ethik in der Medizin* 18(2): 133–147. doi:10.1007/s00481-006-0433-y.

in der Schmitten, Jürgen, Sonja Rothärmel, Christine Mellert, Stephan Rixen, Bernard J. Hammes, Linda Briggs, Karl Wegscheider, and Georg Marckmann. 2011. A complex regional intervention to implement advance care planning in one town's nursing homes: Protocol of a controlled inter-regional study. *BMC Health Services Research* 11: 14. doi:10.1186/1472-6963-11-14.

Irvine, Rob, Ian Kerridge, John McPhee, and Sonia Freeman. 2002. Interprofessionalism and ethics: Consensus or clash of cultures? *Journal of Interprofessional Care* 16(3): 199–210. doi:10.1080/13561820220146649.

Johns, Jeanine L. 1996. Advance directives and opportunities for nurses. *Journal of Nursing Scholarship* 28(2): 149–153. doi:10.1111/j.1547-5069.1996.tb01208.x.

Johnstone, Megan-Jane. 2009. *Bioethics: A nursing perspective*, 5th ed. Sydney/New York: Churchill Livingstone/Elsevier.

Kwak, Jung, and William E. Haley. 2005. Current research findings on end-of-life decision making among racially or ethnically diverse groups. *The Gerontologist* 45(5): 634–641.

Leach Scully, Jackie. 2008. Moral bodies. Epistemologies of embodiment. In *Naturalized bioethics towards responsible knowing and practice*, ed. Hilde Lindemann, Marian Verkerk, and Margret Urban Walker, 23–41. Cambridge: Cambridge University Press.

Malloy, David C., Thomas Hadjistavropoulos, Elizabeth F. McCarthy, Robin J. Evans, Dwight H. Zakus, Illyeok Park, Yongho Lee, and Jaime Williams. 2009. Culture and organizational climate: Nurses' insights into their relationship with physicians. *Nursing Ethics* 16(6): 719–733. doi:10.1177/0969733009342636.

Malone, Ruth E. 2003. Distal nursing. *Social Science and Medicine* 56(11): 2317–2326.

Maust, Donovan T., David M. Blass, Betty S. Black, and Peter V. Rabins. 2008. Treatment decisions regarding hospitalization and surgery for nursing home residents with advanced dementia: The CareAD study. *International Psychogeriatrics* 20(2): 406–418. doi:10.1017/S1041610207005807.

McHugh, Matthew D., Ann Kutney-Lee, Jeannie P. Cimiotti, Douglas M. Sloane, and Linda H. Aiken. 2011. Nurses' widespread job dissatisfaction, burnout, and frustration with health benefits signal problems for patient care. *Health Affairs* 30(2): 202–210. doi:10.1377/hlthaff.2010.0100.

Meehan, Karen A. 2009. Advance directives: The clinical nurse specialist as a change agent. *Clinical Nurse Specialist* 23(5): 258–264. doi:10.1097/NUR.0b013e3181b20809.

Mezey, Mathy, Melissa M. Bottrell, and Gloria Ramsey. 1996. Advance directives protocol: Nurses helping to protect patient's rights. The NICHE faculty. *Geriatric Nursing* 17(5): 204–209; quiz 210.

Moser, Albine, Rob Houtepen, Cor Spreeuwenberg, and Guy Widdershoven. 2010. Realizing autonomy in responsive relationships. *Medicine, Health Care, and Philosophy* 13(3): 215–223. doi:10.1007/s11019-010-9241-8.

Muramoto, Osamu. 2011. Socially and temporally extended end-of-life decision-making process for dementia patients. *Journal of Medical Ethics* 37(6): 339–343. doi:10.1136/jme.2010.038950.

Nortvedt, Per. 2001. Needs, closeness and responsibilities. An inquiry into some rival moral considerations in nursing care. *Nursing Philosophy* 2(2): 112–121. doi:10.1046/j.1466-769X. 2001.00047.x.

O'Neill, Onora. 2002. *Autonomy and trust in bioethics*. Cambridge: Cambridge University Press.

Oberle, Kathleen, and Dorothy Hughes. 2001. Doctors' and nurses' perceptions of ethical problems in end-of-life decisions. *Journal of Advanced Nursing* 33(6): 707–715.

Pauly, Bernardette, Colleen Varcoe, Janet Storch, and Lorelei Newton. 2009. Registered nurses' perceptions of moral distress and ethical climate. *Nursing Ethics* 16(5): 561–573. doi:10.1177/ 0969733009106649.

Pavlish, Carol, Katherine Brown-Saltzman, Mary Hersh, Marilyn Shirk, and Ann-Marie Rounkle. 2011. Nursing priorities, actions, and regrets for ethical situations in clinical practice. *Journal of Nursing Scholarship* 43(4): 385–395. doi:10.1111/j.1547-5069.2011.01422.x.

Peter, Elizabeth, and Joan Liaschenko. 2004. Perils of proximity: A spatiotemporal analysis of moral distress and moral ambiguity. *Nursing Inquiry* 11(4): 218–225. doi:10.1111/j.1440-1800.2004.00236.x.

Prendergast, Thomas J. 2001. Advance care planning: Pitfalls, progress, promise. *Critical Care Medicine* 29(2 Suppl): N34–N39.

Puntillo, Kathleen A., and Jennifer L. McAdam. 2006. Communication between physicians and nurses as a target for improving end-of-life care in the intensive care unit: Challenges and opportunities for moving forward. *Critical Care Medicine* 34(11 Suppl): S332–S340. doi:10.1097/01.CCM.0000237047.31376.28.

Richter, Joerg, and Martin R. Eisemann. 1999. The compliance of doctors and nurses with do-not-resuscitate orders in Germany and Sweden. *Resuscitation* 42(3): 203–209.

Roper, Nancy, Winifred W. Logan, and Alison J. Tierney. 2000. *The Roper-Logan-Tierney model of nursing: Based on activities of living*. Edinburgh: Churchill Livingstone.

Rosnick, Christopher B., and Sandra L. Reynolds. 2003. Thinking ahead: Factors associated with executing advance directives. *Journal of Aging and Health* 15(2): 409–429.

Ryan, Diane, and Mary A. Jezewski. 2012. Knowledge, attitudes, experiences, and confidence of nurses in completing advance directives: A systematic synthesis of three studies. *Journal of Nursing Research* 20(2): 131–141. doi:10.1097/jnr.0b013e318256095f.

Sahm, Stephan, Regina Will, and Gerhard Hommel. 2005. Would they follow what has been laid down? Cancer patients' and healthy controls' views on adherence to advance directives compared to medical staff. *Medicine, Health Care, and Philosophy* 8(3): 297–305. doi:10.1007/s11019-005-2108-8.

Scherer, Yvonne, Mary A. Jezewski, Brian Graves, Yow-Wu B. Wu, and Xiaoyan Bu. 2006. Advance directives and end-of-life decision making: Survey of critical care nurses' knowledge, attitude, and experience. *Critical Care Nurse* 26(4): 30–40.

Schickedanz, Adam D., Dean Schillinger, C.S. Landefeld, Sara J. Knight, Brie A. Williams, and Rebecca L. Sudore. 2009. A clinical framework for improving the advance care planning process: Start with patients' self-identified barriers. *Journal of the American Geriatrics Society* 57(1): 31–39. doi:10.1111/j.1532-5415.2008.02093.x.

Schwerdt, Ruth. 1998. *Eine Ethik für die Altenpflege: Ein transdisziplinärer Versuch aus der Auseinandersetzung mit Peter Singer, Hans Jonas und Martin Buber*, Reihe Pflegewissenschaft. Bern: Huber.

Sellman, Derek. 2011. Professional values and nursing. *Medicine, Health Care, and Philosophy* 14 (2): 203–208. doi:10.1007/s11019-010-9295-7.

Severinsson, Elisabeth. 2003. Moral stress and burnout: Qualitative content analysis. *Nursing & Health Sciences* 5(1): 59–66. doi:10.1046/j.1442-2018.2003.00135.x.

Silveira, Maria J., Scott Y.H. Kim, and Kenneth M. Langa. 2010. Advance directives and outcomes of surrogate decision making before death. *The New England Journal of Medicine* 362(13): 1211–1218. doi:10.1056/NEJMsa0907901.

Stein, Leonhard I., and Madison Wis. 1967. The doctor-nurse game. *Archives of General Psychiatry* 16: 699–703.

166   S. Monteverde

Thompson, Trevor, Rosaline Barbour, and Lisa Schwartz. 2003. Adherence to advance directives in critical care decision making: Vignette study. *BMJ* 327(7422): 1011. doi:10.1136/bmj.327. 7422.1011.

Triplett, Patrick, Betty S. Black, Hilary Phillips, Sarah Richardson Fahrendorf, Jack Schwartz, Andrew F. Angelino, Danielle Anderson, and Peter V. Rabins. 2008. Content of advance directives for individuals with advanced dementia. *Journal of Aging and Health* 20(5): 583–596. doi:10.1177/0898264308317822.

Tschudin, Verena. 2002. *Ethics in nursing: The caring relationship*, 3rd ed. Edinburgh/New York: Butterworth-Heinemann.

Turkel, Marian C., Marilyn A. Ray, and Lynne Kornblatt. 2012. Instead of reconceptualizing the nursing process let's re-name it. *Nursing Science Quarterly* 25(2): 194–198. doi:10.1177/0894318412437946.

Wagner, Pierre-André. 2011. Interdisziplinäre Kooperation zwischen Ethik und Recht. In *Handbuch Pflegeethik. Ethisch denken und handeln in den Praxisfeldern der Pflege*, ed. Settimio Monteverde, 74–82. Stuttgart: Kohlhammer.

Walerius, Theresa, Pamela D. Hill, and Mary A. Anderson. 2009. Nurses' knowledge of advance directives, patient self-determination act, and Illinois advance directive law. *Clinical Nurse Specialist* 23(6): 316–320. doi:10.1097/NUR.0b013e3181be3273.

Wareham, Pauline, Antoinette McCallin, and Kate Diesfeld. 2005. Advance directives: The New Zealand context. *Nursing Ethics* 12(4): 349–359.

Winter, Laraine, Susan M. Parks, and James J. Diamond. 2010. Ask a different question, get a different answer: Why living wills are poor guides to care preferences at the end of life. *Journal of Palliative Medicine* 13(5): 567–572. doi:10.1089/jpm.2009.0311.

Wolf, Zane R., and Patti R. Zuzelo. 2006. "Never again" stories of nurses: Dilemmas in nursing practice. *Qualitative Health Research* 16(9): 1191–1206. doi:10.1177/1049732306292544.

Wright, David, and Susan Brajtman. 2011. Relational and embodied knowing: Nursing ethics within the interprofessional team. *Nursing Ethics* 18(1): 20–30.

Wright, Alexi A., Baohui Zhang, Alaka Ray, Jennifer W. Mack, Elizabeth Trice, Tracy Balboni, Susan L. Mitchell, et al. 2008. Associations between end-of-life discussions, patient mental health, medical care near death, and caregiver bereavement adjustment. *JAMA: Journal of the American Medical Association* 300(14): 1665–1673. doi:10.1001/jama.300.14.1665.

Part IV
Ethical Challenges

Chapter 11
Advance Directives Between Respect for Patient Autonomy and Paternalism

Manuel Trachsel, Christine Mitchell, and Nikola Biller-Andorno

11.1 Introduction

There are two main types of advance directives. One type simply designates a substitute decision-maker, sometimes called a healthcare agent, proxy or surrogate. A more comprehensive advance directive (sometimes called a living will) specifies particular principles or considerations intended to guide action with regard to specific future healthcare decisions and possible medical conditions (Jaworska 2009).

At the time an advance directive is composed, the individual anticipates a future situation in which s/he (1) will have lost decision-making capacity and (2) will be in a condition that requires consent for or refusal of a medical intervention. Currently competent individuals can thus make anticipatory decisions for possible future healthcare situations.

The existence of an advance directive does not necessarily mean, however, that it will be clear to the responsible physician in every case what the patient would have decided. Problems with advance directives include, for instance, vagueness, concerns about authenticity, applicability, the competence of the executor, implausibility, internal contradictions, acceptability, and the suitability of the designated surrogate decision-maker, as well as the question whether the anticipatory decisions are what the patient would actually want now.

Notwithstanding these problems, advance directives are increasingly widely recognized as a legal instrument: in many countries, including the US and most

M. Trachsel (✉) • N. Biller-Andorno
Institute of Biomedical Ethics, University of Zurich, Pestalozzistrasse 24,
CH-8032 Zurich, Switzerland
e-mail: manuel.trachsel@gmail.com; nikola.biller@ibme.ch

C. Mitchell, RN, MS, MTS
Division of Medical Ethics, Harvard Medical School, 641 Huntington Avenue,
Boston, MA 02115, USA
e-mail: christine.mitchell@childrens.harvard.edu

P. Lack et al. (eds.), *Advance Directives*, International Library of Ethics,
Law, and the New Medicine 54, DOI 10.1007/978-94-007-7377-6_11,
© Springer Science+Business Media Dordrecht 2014

Western European states, the wishes expressed in an advance directive have to be respected regardless of the type and stage of disease (Vollmann 2012), unless the directive is legally invalid. However, patients have no *claim right*—i.e. they have no right to demand particular treatments, especially when these are expected to be futile (see e.g. Engelhardt 1989). Instead, patients have the right to consent to or refuse a particular recommended treatment, since every medical treatment represents an intrusion into a person's physical and mental integrity and therefore requires consent.

In many cases, a more or less broad range of interpretation is needed with regard to the meaning and implementation of an individual's healthcare decisions made in advance of their illness. This interpretative process is guided by a number of legal standards and ethical criteria, designed to avoid the traps of paternalism and neglect of autonomy.

11.2   Between Respect for Autonomy and Paternalism

In cases where decision-making incapacity is diagnosed, two situations can be broadly distinguished: either an advance directive is on hand or no written[1] advance directive is on hand.

11.2.1   Advance Directive on Hand

Advance directives are designed to ensure that individual wishes expressed when the person was competent to do so are still respected in the event of decision-making incapacity. Ideally, the wishes formulated in the advance directive are in accordance with the patient's current best interests. However, the wishes expressed in the advance directive may sometimes be regarded as contrary to the incompetent patient's well-being.

According to Olick (2001), advance directives reflect "critical interests" with regard to personal dignity and well-being. Therefore, they have to be respected even if they conflict with current sensations of pleasure and pain. In this case, *respect for autonomy*—one of the four bioethical principles advocated by Beauchamp and Childress (2001)—is given more weight than the principle of *beneficence*. One

---

[1] Verbally expressed wishes are often taken into account in exploring the presumed wishes of the patient. However, they are clearly less authoritative than a properly executed written document. In the US, medical orders for life-sustaining treatment (MOLST) are treated like advance directives even though they are not initiated by the patient; they merely record the healthcare provider's conversation with the patient in the form of an order kept in the patient's medical record and applicable across various healthcare locations, such as hospitals, nursing homes, ambulances, hospices and the patient's home.

example would be a patient's wish, expressed in an advance directive, not to receive pain medication that could impair consciousness. Now, the patient, suffering from end-stage cancer, is in a palliative situation in which only opioids could provide significant pain relief. According to the advance directive, the physician is not supposed to administer opioids, no matter how excruciating the patient's pain may be.

11.2.2   No Advance Directive on Hand

For patients who have not prepared an advance directive, treatment decisions are made by surrogates such as family members (see e.g. Zellweger et al. 2008). Under such circumstances, the principle of *beneficence* may sometimes be given greater weight than *respect for autonomy*, as in the following case.

An otherwise happy elderly person with multiple chronic conditions and decision-making incapacity has temporary kidney failure that could be reversed with dialysis. The patient does not have an advance directive, but when still competent she stated repeatedly to family members and medical care providers that she would not wish to be "dependent on machines" to continue living. Nevertheless, in this case, the responsible physician—having consulted the patient's relatives, who see this as a temporary health crisis in an otherwise stable health situation with an apparently fair quality of life—decides to treat the patient's kidney failure.

Tensions between respect for autonomy and beneficence frequently arise, whether or not a patient has an advance directive. In attempting to resolve such tensions, healthcare providers may err on the side of paternalism or on the side of unwarranted respect for supposedly autonomous decisions which do not in fact reflect competent choices.

11.2.3   Paternalism

Paternalism can be defined as "the interference of a state or an individual with another person, against their will, and defended or motivated by a claim that the person interfered with will be better off or protected from harm" (Dworkin 2010). According to this definition, paternalism always involves a certain degree of constraint on a person's freedom or autonomy for particular reasons. The following two examples illustrate paternalistic behaviour:

1. Out of compassion, a forensic physician tells the parents of a victim of violence that their daughter died instantly, whereas in fact she suffered a dreadful death.
2. The wife of an alcoholic hides her husband's liquor bottles because she is worried about his health.

Paternalistic behaviour may be characterized as weak (soft) or strong. According to *weak paternalism*, "a man can rightly be prevented from harming himself (when other interests are not directly involved) only if his intended action is substantially nonvoluntary or can be presumed to be so in the absence of evidence to the contrary" (Feinberg 1971). *Strong paternalism* is embraced when a person is protected "against his will, from the harmful consequences even of his fully voluntary choices and undertakings" (Feinberg 1971).

An example of weak paternalism is the situation in which a patient specifies in his advance directive a desire to continue taking some sort of complementary medication; his physician, however, discovers that the medication causes significant harm to the patient, which she presumes the patient was not aware of. As she can no longer discuss this with the patient, who is now incompetent, she overrides the patient's advance directive, stopping the treatment for the patient's benefit.

An example of strong paternalism is a case where a patient whose valid advance directive clearly states that he refuses hospitalization for any medical reason is hospitalized overnight to receive intravenous hydration for life-threatening dehydration.

The motivation for potentially justifiable—weak or strong—paternalism is usually the desire to avoid harm (non-maleficence) and/or to benefit the person whose autonomy is overridden or compromised.

One could simply argue that, in sum, paternalistic behaviour probably produces more good than harm. But is this really true? According to Gerald Dworkin (2010), this largely depends on our understanding of the good. If the good simply comprises longer life, better health or relief from pain, paternalism might well be an effective strategy. However, for many people, the good also includes elements such as the right to make self-guided decisions. While paternalism can be considered an acceptable moral stance when autonomy is absent or at least in doubt, overriding an individual's explicit, autonomous choice for the sake of promoting his or her well-being is difficult to justify morally.

11.2.4 Respect for Autonomy

Autonomy or *self-determination* is a person's ability to make his or her own self-guided decisions. The principle of *respect for autonomy* obligates healthcare professionals to honour competent patients' informed, voluntary decisions.

According to Ronald Dworkin (1993), a person with the capacity for autonomy needs (1) the ability to espouse a "genuine preference or character or conviction or a sense of self", which could be called the *ability to value*, and (2) the ability to act out of one's sense of conviction, which Jaworska (2009) calls "the ability to enact one's values in the complex circumstances of the real world". These crucial abilities are missing in many disorders, such as severe dementia or loss of consciousness.

If it is possible to apply a specific advance directive directly to a given situation, a *conflict between respect for autonomy and paternalism* may not occur. In this case, the expressed wishes can be transformed into action without restriction.

However, the conflict becomes relevant if an advance directive is formulated vaguely or cannot be directly applied to the present medical situation. In this more difficult case, the advance directive can only serve as a decision aid or a source for inferring the patient's presumed wishes. For example, if an advance directive contains a detailed statement of treatment preferences for end-stage cancer, this statement is not necessarily useful if the patient suffers not from cancer but from end-stage liver cirrhosis with hepatic encephalopathy and loss of consciousness. This example refers to the *accuracy of fit* that is part of the *validity* of advance directives (see Sect. 11.3).

But even if choices are clearly expressed and obviously apply to a specific situation, the range of choices that need to be respected is not unlimited: for example, certain preferences would impose an undue risk or burden on others, costly but futile interventions would place an unjustifiable burden on a limited public healthcare budget, and refusal of basic hygiene might be intolerable for those who care for the patient. The exact scope of what can be claimed or refused is controversial. Disagreements over what wishes need to be respected and what one person can legitimately ask of another are illustrated by the case of active euthanasia.

11.3 Legal Standards and Ethical Criteria for Assessing the Validity of Advance Directives

11.3.1 Legal Standards

In most countries, legal standards for a valid advance directive require a written form that is personally signed by a person who is of age (legal majority), has decision-making capacity, is informed about the decision to be taken (including alternatives to the chosen action), and is able to make and communicate a free (uncoerced) choice.

At the time of composing an advance directive, a person is required to have *decision-making capacity*. The following criteria are typically used for medical decision-making capacity: (1) ability to understand the relevant information, (2) ability to appreciate the medical consequences of the situation, (3) ability to reason about treatment choices, and (4) ability to communicate a choice (Grisso and Appelbaum 1998). Criteria may differ slightly from country to country, but the basic concept is the same (see e.g. Swiss Academy of Medical Sciences 2013). A variety of instruments aid the assessment of decision-making capacity (Sessums et al. 2011).

Decision-making incapacity is caused by a broad range of clinical conditions, such as loss of consciousness due to severe somatic illness, dementia (e.g. Alzheimer's disease or Lewy body disease), brain injury and psychiatric diseases (e.g. schizophrenia or severe depression).

It is especially difficult to assess *retrospectively* whether a patient had decision-making capacity at the time he or she composed an advance directive. Frequently, a patient diagnosed as incompetent to make a particular healthcare decision has an advance directive that was written many years ago. If, for instance, a patient suffers from slow progressive dementia, it can be difficult to establish whether the person was still competent 5 years ago when he or she wrote the advance directive. The ethical criteria presented below can be used to test the moral appropriateness of heeding the contents of an advance directive. In addition, it may be helpful to interview relatives, friends, physicians and other care professionals who have been in contact with the person over a longer period.

Free choice means that a person composing an advance directive has to be able to make an autonomous decision and to communicate the choice without feeling threatened, under duress or external pressure. Ideally, the living will originates from a person's idiosyncratic substrate of wishes and values. According to Beauchamp and Childress (2001), three conditions constitute an *autonomous decision*: (1) the act was carried out intentionally, (2) the act was carried out with an understanding of the important facts and circumstances and (3) the act was carried out without external "controlling influences".

If these legal standards are not met, the advance directive cannot be used to justify medical decisions. If the legal standards are met, an analysis based on ethical criteria can follow.

11.3.2   Ethical Criteria

When an advance directive is formulated vaguely or cannot be directly applied to the present medical situation, criteria are needed to judge its ethical validity, helping to prevent unwarranted paternalism or undue respect for autonomy in cases where there was no competent choice. Four main characteristics have been proposed as *ethical criteria* for assessing the validity of advance directives (see also Trachsel et al. 2013):

1. accuracy of fit
2. plausibility/authenticity
3. lack of contradictions
4. coherent value system

Accuracy of fit means that the clinical situation in question corresponds to the situation envisaged in the advance directive. This does not necessarily imply that advance directives have to be overly specific, as it may be difficult or impossible for the patient to fully anticipate the details of their diagnoses and prognoses, and to make an informed choice based on an appreciation of the options available. However, it is certainly helpful for the patient, family members and the health professionals concerned if the patient's preferences and values are clearly stated, as well as any particular wishes about interventions such as blood transfusions or mechanical life support in the form of ventilators, artificial hearts, etc.

An advance directive is *plausible* and *authentic* when it is in accordance with one's distinctive wishes, personality, character and lifestyle. For relatives and physicians who know the patient, an advance directive will be easiest to accept as representative of the patient's wishes when the content is consonant with his or her personal traits.

The concept of *authenticity* has been extensively debated (e.g. Golomb 1995; Wood et al. 2008). According to a widely shared position (Frankfurt 1988; Glannon 2008), persons are authentic if they can identify with their mental states. For example, wishes expressed with regard to pain medication are authentic if they are formulated by a person who has suffered from chronic pain (mental state) for years, and if the person is able to attest through her or his higher-order reflective capacity that this chronic pain is relevant to the wishes specified in the advance directive.

However, authenticity is not a legal requirement for the validity of advance directives, and it is contentious as an ethical criterion (Brauer 2008). Legally, a person is free to refuse a certain treatment regardless of his or her reasons and even in the absence of particular reasons. Accordingly, Olick (2001) states that an advance directive is not required to be an authentic expression of its author. The requirement of authenticity would open the floodgates to paternalistic actions, as it would be quite easy to evaluate an advance directive as non-authentic and non-autonomous. Instead, it is sufficient to see an advance directive as an "intentional plan to assert control over one's dying process" (Olick 2001).

It seems self-evident that an advance directive should not contain internal *contradictions* or contradictory instructions with regard to one and the same medical situation. For instance, a patient's living will cannot be honoured when one part of the advance directive refuses withdrawal of treatment in every imaginable situation and requests that everything possible be done to obtain a lung transplant, while another part of the same advance directive requests withdrawal of treatment in end-stage cystic fibrosis.

The ethical validity of an advance directive is more obvious when the wishes expressed are evidently based on a *coherent value system*. This can be defined as a set of values that are interconnected in a logical and hierarchical manner and that guide a person's preferences, decisions and actions. The value system need not be highly complex and abstract, nor does the absence of an identifiable coherent value system render an advance directive invalid. In fact, it is controversial whether and how ethical values should be communicated to healthcare professionals and relatives via an advance directive at all (Brauer 2008).

The ethical criteria described above can provide important guidance in assessing the validity of advance directives that are, for instance, formulated vaguely or cannot be directly applied to the clinical situation. However, it is important to discuss the precise role of these criteria in the assessment. Some, such as accuracy of fit and lack of contradiction, are fairly uncontroversial as a matter of principle but may be applied more or less strictly. Others, such as authenticity and a coherent value system, remain controversial as regards both interpretation and appropriateness. Even so, they capture important aspects of the debate on advance directives and can help to articulate the reasons for moral disagreement.

The criteria discussed in this section focus on the choices expressed by a rational individual moral agent. However, the situations advance directives aim to anticipate are likely to be characterized to a great extent by dependence on others. It is thus of interest to explore what a relational perspective can add to the discussion on advance directives.

11.4   Advance Directives and Relationships: The Ethics of Care Perspective

The fundamental conflict between *respect for autonomy* and *paternalism* is part of every social relationship. Alongside other ethical approaches, the ethics of care (Held 2005) provides an important theoretical perspective on this conflict.

The *ethics of care* is a form of *relational ethics* in the sense that "its central focus is on the compelling moral salience of attending to and meeting the needs of the particular others for whom we take responsibility" (Held 2005). The ethics of care respects the fact that persons depend on others for most of their lives. The ethics of care "addresses rather than neglects moral issues arising in relations among the unequal and dependent, relations that are often emotion-laden and involuntary" (Held 2005). The family context is prototypical for such relations.

Degrees of dependence may vary over the life course; for instance, children or persons in situations of illness or after accidents will need a lot of care. But even healthy adults are likely not to be completely self-sufficient, but need others even for their everyday professional and private activities. Later in life, many people need care every day, and some individuals with disabilities may be dependent on care throughout their lives.

Most people composing an advance directive do so with a view to a future situation of involuntary dependence in which they need the care of others. Focusing on individual preferences and trying to extend individual autonomy may not do justice to the challenges posed by this new state of significant need and dependence.

On the other hand, advance directives are not necessarily antithetical to a care perspective. The ethics of care does not postulate that there is no room for private decisions that may also go against the expectations or wishes of close persons. Advance directives can specify the relational network in which the individual is situated and highlight trustful relationships. Also, advance directives need not be a vote of no confidence in the treating physicians or caring relatives; they may even serve as an "icebreaker", making it easier for healthcare professionals and relatives to communicate about the patient's preferences and interests. Not surprisingly, a randomized controlled study found that advance care planning including the formulation of an advance directive "improves end of life care and patient and family satisfaction and reduces stress, anxiety, and depression in surviving relatives" (Detering et al. 2010).

Furthermore, the ethics of care values sympathy, antipathy, anger, responsiveness or other feelings as important *moral emotions* that should guide behaviour no less than rational arguments. This puts a new complexion on the conflict between respect for autonomy and paternalism. Even if an advance directive is not fully consistent and rational, this does not mean that it is completely irrelevant and that the only option is to override it in a paternalistic manner. Instead, it is advisable to place more reliance on the emotions expressed in the document, which can provide an important basis for discussing the implementation of an advance directive.

11.5   Consistency in the Implementation of Advance Directives

Ethical criteria for assessing validity need to be calibrated in such a way as to strike a balance between paternalism and a form of consumerism that would let patients have their way even if their advance directive is not an expression of a competent choice. Even though some of the requirements (e.g. for a coherent value system) may be controversial, measuring individual advance directives against these ethical criteria can help to promote consistent implementation among physicians and healthcare teams. Beyond individual judgements, the ethical criteria also provide a framework for discussing consistent implementation of advance directives within medical communities (e.g. groups of providers or medical subdisciplines).

The requirement for consistency is fairly straightforward: if two similar patients with similar health problems compose similar advance directives, the patients should be treated similarly. If the two patients, their health problems or their advance directives differ in essential respects, it is perhaps not appropriate to treat the two patients similarly. Yet this claim raises a lot of questions. Should patients with decision-making incapacity who have the same disease (e.g. end-stage brain cancer) and a very similar advance directive be treated similarly, even if one patient is 30 and the other 90 years old? Perhaps both have stated in their advance directive that they do not wish to receive further surgical treatment for their cancer once they become incapable of decision-making. Intuitively, one may be more inclined to accept this living will if the patient is 90 because of the whole life span we could imagine ahead of the 30-year-old patient. Yet this would constitute an age bias that is not part of the advance directive concept. An advance directive is valid regardless of the patient's age. For instance, even a child of 10 years can have decision-making capacity with regard to some vitally important decisions.

There may be other sources of potential bias: physicians may be more inclined to implement an advance directive if they agree with the wishes expressed by the patient. Thresholds for the validity of an advance directive might be raised when physicians completely disagree with the content of an advance directive, particularly with regard to morally highly charged issues such as assisted suicide. Economic factors might also influence the acceptance of an advance directive. Relatives might, for instance, not want to let go of their loved one and argue for a very strict interpretation of standards; conversely, they might be worried about the costs accumulating for the care of their relative, whose quality of life they consider to be very poor. The requirement of consistency calls for a given advance directive to be interpreted in the same way regardless of biasing factors.

11.6   Conclusions

In cases where decision-making incapacity is diagnosed, the existence of an advance directive does not necessarily mean that it will be clear to the responsible physician in every case what the patient would have decided. Problems with advance directives include vagueness, the question of authenticity, applicability, the competence of the executor, implausibility, internal contradictions, acceptability, or the question whether the anticipatory decisions are what the patient would actually want now.

Because advance directives are not always clearly formulated, a certain degree of interpretation is demanded of the responsible physician. In interpreting advance directives, healthcare professionals find themselves in an area of conflict between respect for autonomy, on the one hand, and paternalism on the other.

Besides legal requirements, it is important to apply ethical criteria—including accuracy of fit, plausibility/authenticity, lack of contradictions and a coherent value system—for assessing the validity of advance directives, although there is certainly room for discussion as to the specific requirements which these criteria should entail.

The fundamental conflict between respect for autonomy and paternalism is part of every social relationship. Alongside other ethical approaches, the ethics of care (Held 2005, 2006) provides an important theoretical perspective on this conflict. Advance directives are composed for a future situation of involuntary dependence, in which someone needs the care of others. Advance directives are not a vote of no confidence and could even ease the burden on close relationships, serving as critical icebreakers for communication between patients, relatives and healthcare professionals regarding the care patients receive when they are no longer able to speak for themselves.

References

Beauchamp, Tom L., and James F. Childress. 2001. *Principles of biomedical ethics*, 5th ed. New York: Oxford University Press.
Brauer, Susanne. 2008. Die Autonomiekonzeption in Patientenverfügungen – Die Rolle von Persönlichkeit und sozialen Beziehungen. *Ethik in der Medizin* 20(3): 230–239.

Detering, Karen M., Andrew D. Hancock, Michael C. Reade, and William Silvester. 2010. The impact of advance care planning on end of life care in elderly patients: Randomised controlled trial. *BMJ* 340: c1345.

Dworkin, Ronald. 1993. *Life's dominion: An argument about abortion, euthanasia, and individual freedom*, 1st ed. New York: Knopf.

Dworkin, Gerald. 2010. Paternalism. In *The Stanford encyclopedia of philosophy* (Summer 2010 edition), ed. Edward N. Talta. http://plato.stanford.edu/entries/paternalism/

Engelhardt Jr., H. Tristram. 1989. Advance directives and the right to be left alone. In *Advance directives in medicine*, ed. Chris Hackler, Ray Moseley, and Dorothy E. Vawter, 141–154. New York: Praeger.

Feinberg, Joel. 1971. Legal paternalism. *Canadian Journal of Philosophy* 1(1): 105–124.

Frankfurt, Harry G. 1988. Identification and wholeheartedness. In *The importance of what we care about*, ed. Harry G. Frankfurt. New York: Cambridge University Press.

Glannon, Walter. 2008. Psychopharmacological enhancement. *Neuroethics* 1(1): 45–54.

Golomb, Jacob. 1995. *In search of authenticity: From Kierkegaard to Camus*. London: Routledge.

Grisso, Thomas, and Paul S. Appelbaum. 1998. *Assessing competence to consent to treatment: A guide for physicians and other health professionals*. New York: Oxford University Press.

Held, Virginia. 2005. Ethics of care. In *The Oxford handbook of ethical theory*, ed. David Copp. New York: Oxford University Press.

Held, Virginia. 2006. *The ethics of care: Personal, political, global*. Oxford: Oxford University Press.

Jaworska, Agnieszka. 2009. Advance directives and substitute decision-making. In *The Stanford encyclopedia of philosophy* (Summer 2010 edition), ed. Edward N. Talta. http://plato.stanford.edu/entries/advance-directives/

Olick, Robert S. 2001. *Taking advance directives seriously: Prospective autonomy and decisions near the end of life*. Washington, DC: Georgetown University Press.

Sessums, Laura L., Hanna Zembrzuska, and Jeffrey L. Jackson. 2011. Does this patient have medical decision-making capacity? *JAMA: The Journal of the American Medical Association* 306(4): 420–427.

Swiss Academy of Medical Sciences. 2013. *End-of-life care*. Medical-ethical guidelines. Basel: SAMS.

Trachsel, Manuel, Christine Mitchell, and Nikola Biller-Andorno. 2013. Decision-making incapacity at the end of life: Conceptual and ethical challenges. *Bioethica Forum* 6(1): 26–30.

Vollmann, Jochen. 2012. Patientenverfügungen von Menschen mit psychischen Störungen. Gültigkeit, Reichweite, Wirksamkeitsvoraussetzung und klinische Umsetzung. [Advance directives in patients with mental disorders. Scope, prerequisites for validity, and clinical implementation]. *Der Nervenarzt* 83(1): 25–30.

Wood, Alex M., P. Alex Linley, John Maltby, Michael Baliousis, and Stephen Joseph. 2008. The authentic personality: A theoretical and empirical conceptualization and the development of the Authenticity Scale. *Journal of Counseling Psychology* 55(3): 385–399.

Zellweger, Caroline, Susanne Brauer, Christopher Geth, and Nikola Biller-Andorno. 2008. Advance directives as an expression of individualistic autonomy? The role of relatives in advance directive forms. *Ethik in der Medizin* 20(3): 201–212.

Chapter 12
The Use of Advance Directives in the Context of Limited Resources for Healthcare

Ruth Horn and Ruud ter Meulen

12.1 The High Costs of Dying

Since the nineteenth century, medical care has seen continuous technological advances enabling physicians to make a reliable diagnosis for all kinds of disease, to cure many diseases in a safe and effective manner, and to support public health measures such as infection control and screening for specific diseases. Technological advances allow diseases and other health problems that could not be cured in the past to be treated effectively today (Jones 2002). Medical technology is nowadays hugely admired, and there is a strong expectation that it will be able to heal all our diseases and even to extend our lives indefinitely.

But there are concerns, firstly about the power of medical technology—particularly, the power to put itself at the heart of medical care at the expense of other practices and moral values. Certainly in the final stages of life, there is a concern that medical technology takes control of the dying process, while the patient might be much better off in a hospice setting, for example. Instead of heroic measures to save and extend life, a hospice offers a caring environment in which pain is alleviated, personal relationships are established and meaning can be found. All too often, such caring practices have to give way because of expectations or fascination about what technology might be able to do to get the patient "back to life". As a result, patients often die in a hospitalized environment, surrounded by bleeping machines and by a family frightened by what is happening to their loved one and feeling out of control. The patients themselves have lost the sense of what is going on and have lost the capacity to make decisions about their care.

R. Horn, Ph.D. (✉)
The Ethox Centre, University of Oxford
e-mail: ruth.horn@ethox.ox.ac.uk

R. ter Meulen, Ph.D.
Centre for Ethics in Medicine, University of Bristol, Bristol, UK
e-mail: r.terMeulen@bristol.ac.uk

P. Lack et al. (eds.), *Advance Directives*, International Library of Ethics, Law, and the New Medicine 54, DOI 10.1007/978-94-007-7377-6_12,

A second problem has to do with costs. Medical technology is usually expensive and is considered to be the most important factor in the rise of healthcare costs. Newhouse (1992) concludes that more than 50 % of the rise in healthcare costs in recent decades has been caused by technological change, "or what loosely might be called the march of science and the increased capabilities of medicine". According to Jones (2002), healthcare expenditures across the OECD increased between 1960 and 1997 from about 4 % to about 8 % of GDP on average. At least half, and most likely three quarters or more, of this change seems to be driven by medical technology. The majority of costs are incurred in the final year of life. Researchers in Manitoba, Canada, for example, found that 1.1 % of the population consumed 21.3 % of healthcare costs and that these costs arose in the final 6 months of life (Fassbender et al. 2009). In the US, just 10 % of the 24 million Medicare beneficiaries who received inpatient or outpatient hospital care in 2009 accounted for 64 % of the costs of the programme. Care associated with the end of life accounts for 10–12 % of the overall US health budget and 27 % of the Medicare budget (Lubitz and Riley 1993). Medicare provides access to medical care for people over 65 and for people with disabilities. As Callahan (2009: 1) points out, healthcare costs have increased annually by between 7 and 12 % for many years and the rate is expected to remain at around 6–7 %. In 2007, 16.3 % of US GDP was spent on healthcare, a figure expected to rise to 19.5 % in 2017. Medicare's budget is projected to climb from $427 billion in 2007 to $884 billion in 2017. As Medicare accounts for a substantial part of the US health budget, the overall costs of care will rise correspondingly. According to Callahan (2009: 2):

> The projections are that the Medicare program will go bankrupt in 10 years or so, and the overall cost of health care will rise from $2.2 trillion in 2007 to over $4 trillion in the next decade or so, an astonishing jump.

The reasons why attempts to cut the costs seem to fail have to do with the increasing availability of technologies able to treat ever more pathologies and an increase in the number of those who use them (Callahan 2009: 23).

Though Medicare does fund the treatment of older people (over 65), there is increasing evidence that the high costs of care in the final year of life cannot be attributed to age per se; rather, the costs of care are more related to proximity to death. People close to death have much higher health expenditures than those at the same age who survive (Miller et al. 2004). According to Zweifel et al. (1999), average healthcare expenditures are higher for elderly than for non-elderly persons not because of higher morbidity but because of higher mortality. Further evidence in support of the hypothesis that healthcare expenditures depend primarily on time to death rather than age is provided by McGrail et al. (2000) and Yang et al. (2003). Elderly persons consume much more healthcare in the last year of life no matter how old they are when they die. Research in the UK found a similar pattern for inpatient care: irrespective of age, patients tended to use most of their lifetime bed days in the year immediately before death (Dixon et al. 2004). The results of the UK study confirm that "the highest proportion of costs for acute care are incurred in the final years of life, no matter at which age this happens to be". According to the researchers, the "total costs of acute care are greater in elderly people simply because this age group makes up a larger proportion of dying people." As we will discuss in the following section, the benefits of high-cost acute care for this population are questionable, particularly with regard to quality of life and physical welfare in the sense of respecting patients' preferences. It is from this perspective that advance directives become relevant means to put technology on hold and to help patients take control of end-of-life decision-making, even if they are no longer able to express their wishes. In the next section, we seek to answer the question whether advance directives can indeed lead to reductions in the costs of care and under what conditions such a use of advance directives can be justified.

12.2   Medicalization of Old Age

The elderly nowadays receive care that was unimaginable two or three decades ago. Owing to new technological developments, the elderly are treated at an increasing age with sophisticated medical technologies for acute medical conditions. Since the early 1990s, the number of people aged over 80 or even 90 receiving open heart surgery, organ transplantation or renal dialysis has increased sharply (Natarajan et al. 2007; Turrentine et al. 2006). Though such treatments can be beneficial and enhance quality of life, some of these technologies are less successful. This is particularly true of intensive care technology, which is increasingly accessed by old and very old patients. A study by Bagshaw et al. (2009) in intensive care units (ICUs) in Australia and New Zealand found out that the proportion of patients aged over 80 was rapidly increasing. However, age $\geq$ 80 years was associated with higher ICU and hospital death compared with younger age strata. Factors associated with lower survival rates included admission from a chronic care facility, comorbidity, non-surgical admission and longer stay in the ICU. Patients who survived (80 %) usually did not return home but were discharged to nursing home care or other long-term care facilities. A study by Roch et al. (2011) on survival factors for 299 patients aged 80 years or over admitted to intensive care between 2001 and 2006 found that hospital mortality was 55 %. Up to 50 % of the patients discharged from hospital were still alive at 2 years, but mortality was three times higher than for the general population. Mortality was particularly high among patients with severe acute disease at admission and also after discharge.

Quality of life after discharge to a nursing home is often much lower than for the reference population. Cuthbertson et al. (2010) followed 300 patients with median age > 60 for a period of 5 years after ICU discharge. They found that ICU admission is associated with high mortality, poor physical quality of life and low quality-adjusted life years gained compared to the general population. In a study of 1-year trajectories after ICU discharge, Unroe et al. (2010) found that patients with a poor outcome were older, had more comorbid conditions and were more frequently discharged while still receiving mechanical ventilation than those with better outcomes. They were particularly concerned about patients with intermediate outcomes, i.e. those who are alive but have moderate functional dependency. These previously high-functioning patients rarely improved over time; instead, they were cycling frequently between post-acute care facilities and hospitals. The authors argued that prolonged mechanical ventilation is a highly resource-intensive condition with a generally poor outcome. Given the disproportionately high costs and profound disability associated with prolonged mechanical ventilation, the authors advised clinicians to reconsider their approach to its provision. They also recommended that physicians should not only discuss long-term outcomes with surrogates in terms that they can easily understand but also explicitly convey the probable burdens of treatment and the likelihood of future functional dependency.

In elderly patients, prolonged mechanical ventilation and use of other intensive care technologies can be signs of overtreatment. According to Jennett (1995), patients of all ages are frequently overtreated, particularly those who present with critical illness, requiring emergency surgery, resuscitation or intensive care. Jennett observes that the most common reason for overtreatment is ignorance about the probability of benefit for a patient of a particular age and severity of illness. This is particularly true for the elderly because there is not much evidence about their response to critical care, as they are often excluded from clinical trials. But even if data are available, treatment decisions are often influenced by pressures from colleagues, nurses or relatives who expect interventions to be performed or who are keen to do anything to save the life of a beloved family member. Such a situation may lead to a cascade of treatments which is difficult to stop and which may result in substantial costs. The case reported by Lisa Krieger (2012), who witnessed the death of her father, is an illustration of such a process. It is summarized in an interview with Krieger conducted by Daniel Callahan (2012):

> As Krieger put it, "The medical nightmare started, as they so often do, incrementally." While her father had made clear prior to his dementia that he wanted to die a "natural death," what he got ... was an unnatural "death by medicine," as someone once put it. The total cost for the hospital stay alone was $323,000. Again and again Krieger had to make a decision about going on, as one crisis after another surfaced. With each new crisis, the doctors offered hope. There was, they said, "a decent chance we could turn it around." They could not, and he finally died. But as the days moved along from one crisis after another, Krieger kept asking herself, Was it all worth it? "Should we have quit?" she wrote. And when?

12.3   Modern Medicine at the End of Life

The case of Lisa Krieger's father must be situated in the context of the US, where there are many financial and professional incentives for physicians to continue with aggressive but expensive treatments. Instead of accepting that death will come sooner or later, many physicians do not give up the fight to add a few days or months to the life of dying patients—even when, like Lisa's father, they have advance directives in place.

In his book *Setting Limits. Medical Goals in an Aging Society*, Daniel Callahan (1987) argues that we should set limits to the use of expensive medical technology which extends the lives of elderly people for only a few weeks or months. Instead of extending life, medicine should put more efforts into enhancing life: human beings are going to die sometime, in spite of all efforts to extend life. Life cannot be extended indefinitely: this is not possible, nor is it desirable, as a longer life will result in worsening health. In this book, Callahan proposes that intensive medical treatment for the elderly be limited, making such treatments more available for the young, in order to increase their chances of reaching old age. According to Callahan, we all have the moral intuition that to die young is a tragedy, but that to die at an old age is "only a sadness". People who die young have not had the chance to realize their opportunities; their lives are ended prematurely. On the other hand, old people have had enough opportunities to lead a meaningful life. Their death might be more "tolerable", than the death of a young person. According to Callahan, most of us would opt for a greater chance to reach old age or, what he calls a "natural life span", which is not *biological*, but *biographical*. It is the period in which we have had the opportunity to realize our potential, whereupon death is a sad, but not tragic, event. After this natural life span has been reached, at age 75 or 80, life-extending therapies (which are financed by society) should be limited. This proposal ought to be put into practice on the strict condition that the elderly will be offered improved access to long-term care. Limits to "acute care" must go hand in hand with more and better "comfort care".

Callahan's proposal has been strongly criticized as a kind of "ageist" discrimination (Barry and Bradley 1991; Binstock and Post 1991). Gerontologists and liberal ethicists, in particular, have argued that every age has its own aims and that nobody can determine for another whether their life is completed or their "natural life span" has been reached. There is no reason, they argue, why an old person should value his or her life less than a younger one. When one considers only years of life instead of life itself, one shows no respect for the unique value of the human person, which is the moral basis for our society.

Although Callahan's proposal for a compulsory age limit for (publicly funded) acute medical care is difficult to accept for various (ethical, social and political) reasons, his concerns about the uncritical or unthinking application of life-extending care to vulnerable and dying elderly people are still valid. In his book *Taming the Beloved Beast* (Callahan 2009), he criticizes the Medicare system for rewarding the tendency to pay for continuously increasing costs for diminishing returns in terms of life expectancy. Not only is this development putting the health and social care system in jeopardy, it is not even in the interests of many elderly people who might be better off with palliative care than with aggressive and highly burdensome intensive care technologies.

12.4 Can Advance Directives Help to Cut Costs?

But if age limits are a "no-go", can advance directives help to cut costs while, at the same time, promoting more appropriate care at the end of life? Many elderly patients seem to prefer medical interventions to be limited at the end of life, and instead would like to receive comfort care and palliative treatments (McCarthy et al. 2008). Advance directives to limit aggressive life-extending care would then have the combined result of ensuring comfort care and reducing costs—at least for those who refuse excessive end-of-life treatment.

Since the end of the 1990s, various studies have been conducted to find out whether advance directives indeed result in diminishing costs of care. The first major effort to investigate this question was the US Study to Understand Prognoses and Preferences for Outcomes and Risks of Treatments (The SUPPORT Principal Investigators 1995), which involved 9,105 seriously ill patients. SUPPORT included an observational study (phase I) and a randomized trial with an intervention and a control arm (phase II). The trial aimed to determine whether providing physicians with information about patient prognosis and patient preferences for end-of-life care would have an impact on the use of intensive care units and other hospital resources. The control arm of the study received no specific interventions relating to patient preferences. The intervention arm sought to address physicians' lack of knowledge of patient preferences by giving them information on prognoses and on patient preferences. The intervention also included the involvement of nurses to elicit such preferences and to facilitate advance care planning and enable palliative care. The first part of the study (phase I) documented a wide range of shortcomings in the care provided for seriously ill people. However, the intervention part of the study (phase II) showed that the eliciting and documenting of patient preferences, and communication of this information to physicians, had no impact on the use of hospital resources as compared to the control arm of the study. For example, for the 680 intervention patients who died in hospital, the number of days spent in an ICU, comatose or receiving mechanical ventilation before death, was the same as for the 530 control patients. The authors concluded that:

> The study certainly casts a pall over any claim that, if the health care system is given additional resources for collaborative decision-making in the form of skilled professional time, improvements will occur.

A systematic review by Taylor et al. (1999) of primary studies assessing the association between advance directives and resource use found little evidence to support the hypothesis that advance directives reduce the use of resources by hospitalized patients. Some retrospective studies have shown savings, but these studies had methodological shortcomings. Prospective studies with experimental methods, such as the SUPPORT study, have yielded no evidence of cost savings with the use of advance directives.

Though most studies have failed to provide significant evidence of advance directives contributing to reduced resource use or cost control, a recent study by Nicholas et al. (2011) provided a new perspective by looking into regional differences. The authors argued that advance directives can only influence treatment decisions when the patient's wishes are inconsistent with what would be provided if there was *no* advance directive. They hypothesized that advance directives would have the greatest limiting impact on medical care in regions where the norm is to provide high-level intensive medical interventions at the end of life. They found that advance directives were more common in areas where there were already lower levels of healthcare spending at the end of life. Advance directives had no direct effect on resource use in these areas. However, in areas with high levels of spending on end-of-life care, advance directives had much more impact on hospital use and were associated with a lower likelihood of in-hospital death.

The authors conclude that the clinical impact of advance directives is highly dependent on the regional context in which a patient receives care. They argue that advance directives are important to ensure care consistent with patient preferences in areas where aggressive end-of-life treatment is provided. In areas that already have a lower intensity of care, advance directives will likely have less impact. These findings may explain the sometimes conflicting evidence of the impact of advance directives on end-of-life healthcare expenditures.

Zhang et al. (2009) concluded that increasing communication between patients and their physicians is associated with better outcomes and less expensive care. In this study, 627 advanced cancer patients were interviewed at baseline and followed up through death. Patients were asked specific questions about individual treatment preferences, particularly with regard to end-of-life care. The costs for ICU and hospital stays, hospice care and life-sustaining procedures received in the last week of life were aggregated. Patients who reported having end-of-life conversations with their physicians were less likely to undergo ventilation or resuscitation, or to be admitted to or die in an ICU. They were more likely to receive outpatient hospice care and to stay longer in the hospice. They were in less physical distress than patients without such conversations. The aggregate costs of treatment were 35.7 % lower than for patients without end-of-life discussions. According to Zhang et al., additional analysis showed that higher medical costs in the final week of life were associated with more physical distress in the last week of life and worse overall quality of death as reported by the caregiver.

12.5   Advance Directives and Patient Autonomy

In view of the ever-increasing costs of end-of-life care, there is a serious risk of resources being drawn away from, for example, life-enhancing, chronic or long-term care. The costs of the use of high technology in the final stages of life need to be reduced in order to preserve the right balance of health and social care services, and equitable access to these services. Moreover, intensive care at the end of life is often not effective and not wanted by patients (McCarthy et al. 2008). If advance directives can help to control the costs of care, there are sufficient ethical arguments in favour of their being used for this purpose. However, the evidence about the impact of advance directives on costs is not conclusive. Conversations between physicians and patients, or advance care planning (ACP), seem to be a more effective way to reduce costs, as shown in the study by Zhang et al. (2009).

Advance directives or ACP can only be used to reduce costs if they indeed reflect patients' wishes. Otherwise, they might become blunt rationing instruments to reduce access to services for older people. The ethical debate about advance directives points up various problems relating to their supposed role in enhancing or extending patient autonomy.

One of the main problems associated with advance directives is the question of whether they can express the genuine will of a person in a particular situation. With reference to Buchanan and Brock (1990: 90), who discuss whether it is possible "to view the rights of incompetent individuals as an *extension* of the rights of competent individuals", Davis (2009) distinguishes between what he calls "Extension View" arguments and those which question the moral authority of advance directives. Dworkin (1993: 224), defending the Extension View, argues for example that past preferences should always be respected because of the individual's "right to a life structured by his own values", even if these preferences conflict with the incompetent individual's best interests or (assumed) present wishes. Taking the case of a patient with Alzheimer's disease who formerly stated in an advance directive that she would not want to receive treatments in order to be kept alive in such a condition but now seems to be happy with her life, Dworkin distinguishes between *experiential* interests (quality of life, pleasure, contentment, lack of pain) and *critical* interests (value judgements, basic autonomous decisions). He comes to the conclusion that the latter—determined by the capable individual in accordance with her basic attitude towards life—are more important than the present experiential interests of the incapacitated individual (Dworkin 1993: 201–202). Therefore, the refusal of treatment specified in an advance directive by a person with Alzheimer's disease should be respected, regardless of her currently experienced happiness. Similarly, Buchanan (1988) argues that despite the loss of psychological continuity, personal identity is preserved. Accordingly, advance directives retain their full moral force unless neurological damage is so severe that the living being that remains is not a person any more and has thus no personal identity. Opposing this perspective, Dresser (1994) adopts a critical view of the moral authority of wishes expressed in advance because the competent person may not always be fully informed at the time a decision concerning future treatment is made. In addition, following Parfit (1984), who claims that identity can be reduced to psychological continuity, Dresser (1986) argues that personal identity does not remain the same throughout a life but develops continually, depending on various external and psychological factors. According to her view, an advance directive does not unconditionally express the authentic will of a person in a concrete situation. This argument is backed up by studies which show that preferences for life-sustaining treatment are indeed dependent on the context in which they are made, and that individuals may express different treatment preferences when they are healthy than when they are ill (Ditto et al. 2006).

The difficulty of assessing the authenticity of previously expressed wishes and thus the moral authority of an advance directive surfaces not only in the philosophical but also in the legal debate (Maclean 2006). Whereas countries with a strong emphasis on patient autonomy, such as England, recognize the binding force of advance directives (Horn 2012; Huxtable 2012), specific requirements have to be met so that a directive can be validated. Some authors object that advance directives can in fact be too easily invalidated (Michalowski 2005).

Finally, it appears that advance directives are rarely encountered in practice, and authors such as Fagerlin and Schneider (2004) who first considered that advance directives "serve a strong version of patients' autonomy" come to the conclusion "that living wills do not and cannot achieve that goal". According to these authors, the reasons for this failure are that only few people actually write an advance directive, know what they really want and can articulate their wishes; others are afraid of misinterpretations, or that their living wills will not be taken into consideration.

Given the increasing complexity of medical technologies, how can a patient, particularly an elder person, decide what treatment she wants or does not want to receive once she is incompetent? And how can physicians who have to validate and apply an advance directive be certain that they are acting in accordance with the author's genuine wishes?

To address the moral and practical questions surrounding the use of advance directives, a number of authors (Messinger-Rapport et al. 2009; Halliday 2009; Ozanne et al. 2009) suggest that full and frank physician-patient communication could enable both the physician to better assess the validity of an advance directive and the patient to better express her wishes and preferences.

Studies from the US (e.g. Hammes et al. 2010) show that physician-patient communication on treatment preferences towards the end of life is significantly improved by the use of so-called POLST (Physician Orders for Life-Sustaining Treatments), also known as MOLST/COLST (Medical/Clinician Orders for Life-Sustaining Treatments). These orders, used in more than half of US states (Hickman et al. 2010), have to be completed by the physician together with the patient and/or surrogates and are legally valid medical decisions, which have to be respected when the patient is no longer able to express her wishes. Contrary to advance directives, which can be completed by any (sick or healthy) adult, POLST are employed as a complement to advance directives only for seriously or terminally ill patients. Yet, considering the fact that advance directives are completed mainly by seriously or terminally ill patients (Virmani et al. 1994; Heffner et al. 1996), and that advance directives are often not accessible or wishes are not clearly defined, one could even think of POLST replacing rather than merely complementing advance directives.

Before advance directives are used in the context of cost control, concerns regarding the authenticity of these directives should be taken into account and ACP should take place in a broader sense, in the form of good physician-patient communication about the patient's general but also specific preferences. Such communication could be facilitated by the use of standardized forms such as POLST (Emanuel 2000).

12.6   Conclusion

High-tech, intensive end-of-life care for the elderly leads to increased costs without always improving patients' well-being or corresponding to elderly patients' preferences. In some cases, as we have shown, costly high-tech medical care may even impair the patient's physical and psychological condition. Yet setting limits to intensive medical treatment appears to be difficult for physicians, whose priority is most often to save lives, rather than withdrawing treatment. In order to address these problems, advance directives refusing treatment have been discussed as a means of helping to reduce costs associated with overtreatment or unwanted treatment and thus to better adapt resource use to individual needs and preferences, even when a patient is no longer able to communicate her wishes. Yet there is only conflicting evidence that these directives could have an impact on resource allocation, as physicians seem to be reluctant to respect patients' previously expressed treatment refusals, particularly when such wishes do not accord with the physicians' opinion. These findings must be put in the context of ethical, legal and practical problems regarding the validity and applicability of advance directives. We have argued that the use of advance directives for cost control is only justified if they reflect patients' authentic wishes. However, as discussed above, important questions can be raised as to whether advance directives are valid instruments to express the patient's genuine will in a specific situation and are thus ethically acceptable in the context of cost control and equitable allocation of resources. It is difficult for many patients to anticipate future events and specific preferences, and advance directives are not always known about or taken into account by physicians. Therefore, we argue that the writing of advance directives should not be left to the patient alone. Rather, they should be considered as tools for opening physician-patient communication about difficult issues and individual treatment preferences with regard to end-of-life care. If advance directives were placed in the broader context of ACP, aiming to enhance conversations between physicians and patients, this could be, as pointed out by Zhang et al. (2009), a way forward to increase patient autonomy *and* reduce the costs associated with questionable end-of-life care.

References

Bagshaw, Sean M., Steve A.R. Webb, Anthony Delaney, Carol George, David Pilcher, Graeme K. Hart, and Rinaldo Bellomo. 2009. Very old patients admitted to intensive care in Australia and New Zealand: A multi-centre cohort analysis. *Critical Care* 13(2): R45.

Barry, Robert L., and Gerard V. Bradley (eds.). 1991. *Set no limits: A rebuttal to Daniel Callahan's proposal to limit health care for the elderly.* Urbana: University of Illinois Press.

Binstock, Robert H., and Stephen G. Post. 1991. *Too old for health care? Controversies in medicine, law, economics, and ethics.* Baltimore: The Johns Hopkins University Press.

Buchanan, Allen E. 1988. Advance directives and the personal identity problem. *Philosophy and Public Affairs* 17(4): 277–302.

Buchanan, Allen E., and Dan W. Brock. 1990. *Deciding for others: The ethics of surrogate decision-making.* New York: Cambridge University Press.

Callahan, Daniel. 1987. *Setting limits. Medical goals in an aging society.* New York: Touchstone.

Callahan, Daniel. 2009. *Taming the beloved beast: How medical technology costs are destroying our health care system.* Princeton: Princeton University Press.

Callahan, Daniel. 2012. The trial of "death by medicine": An interview with Lisa Krieger. *Health Care Cost Monitor*. Garrison: The Hasting Center. http://healthcarecostmonitor. thehastingscenter.org/daniel-callahan/the-trial-of-%E2%80%9Cdeath-by-medicine%E2% 80%9D-an-interview-with-lisa-krieger/. Accessed 19 July 2012.

Cuthbertson, Brian H., Siân Roughton, David Jenkinson, Graeme MacLennan, and Luke Vale. 2010. Quality of life in the five years after intensive care: A cohort study. *Critical Care* 14(1): R6.

Davis, John K. 2009. Precedent autonomy, advance directives and end-of-life care. In *The Oxford handbook of bioethics*, ed. Bonnie Steinbock, 349–374. New York: Oxford University Press.

Ditto, Peter H., Jill A. Jacobson, William D. Smucker, Joseph H. Danks, and Angela Fagerlin. 2006. Context changes choices: A prospective study of the effects of hospitalization on life-sustaining treatment preferences. *Medical Decision Making* 26(4): 313–322.

Dixon, Tracey, Mary Shaw, Stephen Frankel, and Shah Ebrahim. 2004. Hospital admissions, age, and death: Retrospective cohort study. *BMJ* 29(7451): 1288.

Dresser, Rebecca. 1986. Life, death, and incompetent patients: Conceptual infirmities and hidden values in the law. *Arizona Law Review* 28(3): 379.

Dresser, Rebecca. 1994. Missing persons: Legal perceptions of incompetent patients. *Rutgers Law Review* 46: 624–630.

Dworkin, Ronald. 1993. *Life's dominion: An argument about abortion, euthanasia, and individual freedom*. New York: Knopf.

Emanuel, Linda. 2000. Living wills can help doctors and patients talk about dying. *BMJ* 320: 1618–1619.

Fagerlin, Angela, and Carl E. Schneider. 2004. Enough: The failure of the living will. *The Hastings Center Report* 34: 30–42.

Fassbender, Konrad, Robin L. Fainsinger, Mary Carson, and Barry A. Finegan. 2009. Cost trajectories at the end of life: The Canadian experience. *Journal of Pain and Symptom Management* 38(1): 75–80.

Halliday, Samantha. 2009. Advance decisions and the Mental Capacity Act. *British Journal of Nursing* 18(11): 697–699.

Hammes, Bernard J., Brenda L. Rooney, and Jacob D. Gundrum. 2010. A comparative, retrospective, observational study of the prevalence, availability, and specificity of advance care plans in a county that implemented an advance care planning microsystem. *Journal of the American Geriatrics Society* 58: 1249–1255. doi:10.1111/j.1532-5415.2010.02956.x.

Heffner, John E., Bonnie Fahy, Lana Hilling, and Celia Barbieri. 1996. Attitudes regarding advance directives among patients in pulmonary rehabilitation. *American Journal of Respiratory and Critical Care Medicine* 154(6): 1735–1740.

Hickman, Susan E., Christine A. Nelson, Nancy A. Perrin, Alvin H. Moss, Bernard J. Hammes, and Susan W. Tolle. 2010. A comparison of methods to communicate treatment preferences in nursing facilities: Traditional practices versus the physician orders for life-sustaining treatment program. *Journal of the American Geriatrics Society* 58: 1241–1248.

Horn, Ruth. 2012. Advance directives in England and France: Different concepts, different values, different societies. *Health Care Analysis*. doi:10.1007/s10728-012-0210-7.

Huxtable, Richard. 2012. *Law, ethics and compromise at the limits of life: To treat or not to treat?* London: Routledge-Cavendish.

Jennett, Bryan. 1995. The elderly and high-technology therapies. In *A world growing old. The coming health care challenges*, ed. Daniel Callahan, Ruud ter Meulen, and Eva Topinkova, 85–96. Washington, DC: Georgetown University Press.

Jones, Charles I. 2002. Why have health care expenditures as a share of GDP risen so much. Working Paper 9325. Cambridge MA: National Bureau of Economic Research.

Krieger, Lisa. 2012. The cost of dying: It's hard to reject care even as costs soar. *Mercury News*. http://www.mercurynews.com/cost-of-dying/ci_19898736. Accessed 19 July 2012.

Lubitz, James D., and Gerald F. Riley. 1993. Trends in Medicare payments in the last year of life. *The New England Journal of Medicine* 328: 1092–1096.

Maclean, Alasdair R. 2006. Advance directives, future selves and decision-making. *Medical Law Review* 14: 291–320.

McCarthy, Ellen, Michael Pencina, Margaret Kelly-Hayes, Jane Evans, Elizabeth Oberacker, Ralph D'Agostino, Risa Burns, and Joanne Murabito. 2008. Advance care planning and health care preferences of community-dwelling elders: The Framingham heart study. *The Journals of Gerontology. Series A, Biological Sciences and Medical Sciences* 63(9): 951–959.

McGrail, K., B. Green, M.L. Barer, R.G. Evans, C. Hertzman, and C. Normand. 2000. Age, costs of acute and long-term care and proximity to death: Evidence for 1987–88 and 1994–95 in British Columbia. *Age and Ageing* 29(3): 249–253.

Messinger-Rapport, Barbara J., Elizabeth E. Baum, and Martin L. Smith. 2009. Advance care planning: Beyond the living will. *Cleveland Clinic Journal of Medicine* 76(5): 287–288.

Michalowski, Sabine. 2005. Advance refusals of life-sustaining medical treatment: The relativity of an absolute right. *Modern Law Review* 68: 958–982.

Miller, Susan C., Orna Intrator, Pedro Gozalo, Jason Roy, Janet Barber, and Vincent Mor. 2004. Government expenditures at the end of life for short- and long-stay nursing home residents: Differences by hospice enrollment status. *Journal of the American Geriatrics Society* 52(8): 1284–1292.

Natarajan, Arun, Samad Samadian, and Stephen Clark. 2007. Coronary artery bypass surgery in elderly people. *Postgraduate Medical Journal* 83(977): 154–158.

Newhouse, Joseph. 1992. Medical care costs: How much welfare loss. *Journal of Economic Perspectives* 6: 3–21.

Nicholas, Lauren H., Kenneth M. Langa, Theodore J. Iwashyna, and David R. Weir. 2011. Regional variation in the association between advance directives and end-of-life Medicare expenditure. *JAMA: The Journal of the American Medical Association* 306(13): 1447–1453.

Ozanne, Elissa M., Ann Partridge, Beverly Moy, Katherine J. Ellis, and Karen R. Sepucha. 2009. Doctor-patient communication about advance directives in metastatic breast cancer. *Journal of Palliative Medicine* 12(6): 547–553.

Parfit, Derek. 1984. *Reasons and persons*. Oxford: Clarendon.

Roch, Antoine, Sandrine Wiramus, Vanessa Pauly, Jean-Marie Forel, Christophe Guervilly, Marc Gainnier, and Laurent Papazian. 2011. Long-term outcome in medical patients aged 80 or over following admission to an intensive care unit. *Critical Care* 15(1): R36.

Taylor, J.S., D.K. Heyland, and S.J. Taylor. 1999. How advance directives affect hospital resource use. Systematic review of the literature. *Canadian Family Physician* 45: 2408–2413.

The SUPPORT Principal Investigators. 1995. A controlled trial to improve care for seriously ill hospitalized patients. The study to understand prognoses and preferences for outcomes and risks of treatments (SUPPORT). *JAMA: The Journal of the American Medical Association* 274 (20): 1596.

Turrentine, Florence E., Hongkun Wang, Virginia B. Simpson, and R. Scott Jones. 2006. Surgical risk factors, morbidity, and mortality in elderly patients. *Journal of the American College of Surgeons* 203(6): 865–877.

Unroe, Mark, Jeremy M. Kahn, Shannon S. Carson, Joseph A. Govert, Tereza Martinu, Shailaja J. Sathy, Alison S. Clay, Jessica Chia, Alice Gray, James A. Tulsky, and Christopher E. Cox. 2010. One-year trajectories of care and resource utilization for recipients of prolonged mechanical ventilation: A cohort study. *Annals of Internal Medicine* 153(3): 167–175.

Virmani, Jaya, Lawrence J. Schneiderman, and Robert M. Kaplan. 1994. Relationship of advance directives to physician-patient communication. *Archives of Internal Medicine* 154(8): 909–913.

Yang, Zhou, Edward C. Norton, and Sally C. Stearns. 2003. Longevity and health care expenditures: The real reasons older people spend more. *The Journals of Gerontology. Series B, Psychological Sciences and Social Sciences* 58(1): S2–S10.

Zhang, Baohui, Alexi A. Wright, Haiden A. Huskamp, Matthew E. Nilsson, Matthew L. Maciejewski, Craig C. Earle, Susan D. Block, Paul K. Maciejewski, and Holly G. Prigerson. 2009. Health care costs in the last week of life: Associations with end of life conversations. *Archives of Internal Medicine* 169(5): 480–488.

Zweifel, Peter, Stefan Felder, and Markus Meiers. 1999. Ageing of population and health care expenditure: A red herring? *Health Economics* 8(6): 485–496.

Chapter 13
From Legal Documents to Patient-Oriented Processes: The Evolution of Advance Care Planning

Tanja Krones and Sohaila Bastami

13.1 Background: The Failure of the Living Will

Over the past few decades, two movements have had an especially high impact on perceptions of end-of-life and palliative care. The development of palliative medicine has promoted treatment goals other than seeking to cure terminally ill patients in their last months of life. The development of biomedical ethics alongside physicians' professional ethics has raised awareness of respect for patients' autonomy as a central value of healthcare.

One of the main results of these scientific, philosophical and societal developments has been the elaboration of ethical, legal and political concepts of advance directives. These are to be used when, as is often the case, patients become incapable of making their own decisions at the end of their lives. By the late 1980s, it was believed that much would be improved if advance directives—which had already been ethically and legally recognized in many countries as important tools for documenting patients' values and wishes—were more widely known and completed. The appointment of surrogate decision-makers was another approach discussed and promoted, sometimes in connection with and sometimes as an alternative to advance directives. Both of these concepts fit into a liberal, treaty-based approach to bioethics, focusing on the promotion of individual autonomy as a key task.

The following case occurred in Switzerland in 2011:

An 83-year-old woman was admitted to the intensive care unit (ICU) after coronary artery stent placement. Prior to myocardial infarction, she had been completely independent. On admission, she was no longer capable of decision-making. Her husband, who had found her unconscious in the garden and called the ambulance in great despair, came to the ICU with an advance directive stating that she had lived her life and did not wish to have

T. Krones (✉) • S. Bastami
Institute of Biomedical Ethics, University of Zurich, Pestalozzistrasse 24,
CH-8032 Zurich, Switzerland
e-mail: tanja.krones@usz.ch

P. Lack et al. (eds.), *Advance Directives*, International Library of Ethics,
Law, and the New Medicine 54, DOI 10.1007/978-94-007-7377-6_13,
© Springer Science+Business Media Dordrecht 2014

life-prolonging treatment in an ICU. The advance directive, in which the husband was appointed as surrogate decision-maker, had been signed one year earlier. The same day, the patient suffered a cardiac arrest and was resuscitated by the ICU team in consultation with the cardiologists in charge. She died in the ICU five days later as a result of her general condition.

This unfortunate story could have happened anywhere in the world. In the US, the landmark Study to Understand Prognoses and Preferences for Outcomes and Risks of Treatments (The SUPPORT Principal Investigators 1995) failed to achieve not only the (secondary) goal of increasing the number of advance directives completed, but also its main goals—improving the quality of care as perceived by healthcare teams, patients and their families, and (most importantly as regards the principle of autonomy) fulfilling the documented wishes of patients through end-of-life treatment decisions. Evidence also indicates that surrogate decision-makers often do not know patients' actual wishes or disagree with their wishes concerning end-of-life care (Shalowitz et al. 2006; Rid and Wendler 2010).

It was against this background that Fagerlin and Schneider (2004) concluded that the living will had "failed". And, given the evidence, this diagnosis was perfectly right. But does this mean that the concepts of advance directives, surrogate decision-making and patient autonomy are misguided, that we should not have shifted from "doctor knows best" to patient autonomy at all? Should strong paternalism be resurrected, at least with regard to end-of-life care, since this is what patients really want and/or what works best?

13.2 Reasons for the Failure of Advance Directives

This is not necessary. Some of the reasons for the failure of the living will are obvious. The situation could be compared to signing an informed consent form for major cardiac surgery without speaking to one's general practitioner, the surgeon or family members. One would wish to know about the procedure, the risks and benefits, and the best or worst possible outcomes. The signed consent forms should be stored and readily available as electronic records, and they should be adhered to. Advance directives comprise informed consent for future life events of existential importance, together with value- and evidence-based care plans which are to be followed. This is a very complex issue, comparable in some ways to successful resuscitation.[1] Resuscitation is not successful in most cases where it is attempted, and this has to do with poor end-of-life care, medical myths and the influence of the media. But consider a case in which resuscitation has a reasonable chance of success—a 40-year-old man in cardiac arrest following a myocardial infarction at a restaurant. If this patient is to be resuscitated, a whole chain of actions and persons closely linked, well equipped and

---

[1] I am grateful to Jürgen in der Schmitten of the German advance care planning project beizeiten begleiten® for this example.

trained to work together has to be in place 24 h a day. Equipment for immediate resuscitation by laypersons has to be available, functional and easy to use. A layperson has to know about the equipment and be able to use it. The person calling the ambulance needs to know the number and what to say, the emergency team has to be available and to come within a given time, and so on.

For a person to write an advance directive on his own and store it in a chest of drawers in his bedroom, without talking to anyone about it, is like calling an emergency number when no pieces of the emergency chain are in place, or some parts are not linked, not well equipped or lacking in knowledge of what to do. The failure of the living will stems from a failure to recognize the relational aspects of autonomy and central human needs for support and communication. In addition, putting an advance directive into practice is a complex undertaking, calling for good planning and a knowledge of how innovations and complex interventions can be effective with regard to patient-oriented outcomes and goals. An advance directive can also be considered as an "intervention" to improve practice, not only as a "right" of patients. Ethics—like advance directives—is not (predominantly) a theoretical or legal matter. "Ethics", as understood in pragmatic philosophy and clinical ethics, "is a practical discipline that deals with real world problems and practices" (Fletcher and Boyle 1997:1). It seeks to provide solutions that promote the ethical principles, beliefs and outcomes we hold dear.

13.3   The Evolution of Advance Care Planning

One of the pioneers of advance care planning is Linda Briggs, co-director of the Respecting Choices® programme at Gundersen Lutheran Medical Center in La Crosse, Wisconsin—one of the most successful programmes for the implementation of advance care planning. Describing the evolution of the field of advance care planning, she writes (Briggs 2003):

> It [. . .] visualizes what can be achieved if advance care planning is transformed from a document-driven, decision-focused event to one that emphasizes a relational, patient-centered process. The outcomes of such a transformation will not be measured by increasing the number of completed advance directives, but by improving satisfaction with the end-of-life experience.

Advance care planning is a new endeavour, which began in the US in the 1990s and is now spreading all over the world. This approach puts the patient and the patient's family at the centre of a well-supported planning and caring process. It facilitates and supports the discussion, management and implementation of future plans with regard to medical and non-medical goals of care. These goals are based on the specific value system of the person concerned. Care plans are evaluated by well-trained medical staff at any relevant point in time, including but not limited to the last weeks of life.

In summarizing the essential aims of advance care planning on its website, the International Society of Advance Care Planning and End of Life Care writes, catchily, of "hoping for the best but planning for the worst" since "failing to plan is planning to fail". This endeavour is also part of a cultural shift towards demystifying death and dying, and replacing "high-tech" medicine, which aims to prolong life by all means, with "high-touch" palliative care, which aims to provide optimal symptom control and support (and may also prolong life).

13.4   Obstacles and Requirements

Best-practice programmes are required, not only when an operation is to be performed, but also to support patients and families in planning ahead when a sudden life-threatening disease occurs, a chronic condition deteriorates, or death is expected during the next 6–12 months. If this planning is to be successful, the necessary structures, persons and skills have to be available. Over the past 15 years, many lessons have been learned from the findings of the SUPPORT study (The SUPPORT Principal Investigators 1995), from reviews of surrogate decision-making and from monitoring patients' wishes about end-of-life care. In particular, the following main obstacles have been identified to achieving the goals of advance directives, understood as informed consent to (or refusal of) future life-sustaining treatments and measures:

1. lack of communication skills among physicians and healthcare teams in discussing end-of-life choices with patients
2. lack of interdisciplinary approaches to end-of-life care communication and documentation
3. for healthcare professionals, lack of information on illness trajectories and evidence-based information on outcomes of life-sustaining end-of-life treatments (e.g. resuscitation, feeding tubes)
4. for patients, lack of understandable, tailored information formats required to achieve the informed consent standard with regard to future decisions
5. lack of well-structured information and shared documentation, both for emergency cases and in general

The following measures are therefore required:

1. implementation of a structured communication process to support a shared understanding among the parties concerned (patients, loved ones, physicians and healthcare teams)
2. development of evidence-based information, tailored to patients' needs, on end-of-life care
3. development of improved documentation of options and wishes discussed, including emergency plans (physician orders for life-sustaining treatment forms), and implementation of structures for sharing of information between institutions

4. design of interdisciplinary educational programmes to overcome existing cognitive and emotional barriers to open interdisciplinary communication in discussing end-of-life choices with patients and their families

Best-practice programmes for advance care planning which incorporate most of these elements include Respecting Choices® (http://respectingchoices.org) and its adaptation to the Singaporean, Australian (www.respectingpatientchoices.org.au) and German context (beizeiten begleiten®: In der Schmitten et al. 2011). Mention should also be made of programmes (e.g. Weiner and Cole 2004) that teach physicians to communicate about physician orders for life-sustaining treatment (POLST) forms, rather than merely using them as legal documents; the UK Gold Standards Framework (www.goldstandardsframework.org.uk) for optimizing care for patients near the end of life; and the advance care planning decision support tools of the Massachusetts General Hospital (Volandes et al. 2009), the Nous Foundation (2009) and Penn State College of Medicine (Schubart et al. 2012).

13.5 Evidence on Advance Care Planning Developments

Evidence in recent years has demonstrated that an open and well-documented communication process, initiated by healthcare professionals with an interdisciplinary approach, is considered most helpful for most patients—both in general and particularly with regard to questions about dealing with future deterioration of health. Almost all patients want to discuss these issues, while many physicians are still reluctant to do so (Back et al. 2007; Baile et al. 2002; Hagerty et al. 2005; Clayton et al. 2007; Stiefel et al. 2010) because of various cognitive and emotional barriers to communication, as well as structural issues (time constraints and reimbursement).

Best-practice advance care planning programmes share several features, including a focus on the promotion of interdisciplinary communication and implementation skills. It has been shown that programmes are most successful if the discussion with the patient and family member(s) is initiated by an informed physician and led by a skilled advance care planning facilitator (non-physician healthcare professional, such as a nurse, social worker or chaplain); the plan should finally be reviewed, documented and, if necessary, put into practice by the physician. Of the utmost importance is a community-based implementation approach, including meta-initiatives on the local and national level (Wilson and Schettle 2012; Silvester 2012) involving professionals (GPs, nursing homes, emergency teams, hospitals and hospices), politicians, media and the public.

In La Crosse, one of the regions where advance care planning has been implemented for several years, a retrospective study on death records in 2006/2007 revealed that 90.0 % of decedents had a written AD at the time of death, the advance directive was available in the medical record in 99.4 % of cases, and in 90 % of cases patients' instructions were documented by the treating physician as physician orders, which were followed in all cases (Hammes et al. 2010).

A randomized controlled trial of the Australian programme based on Respecting Choices further revealed that surviving relatives have significantly lower levels of depression after the death of their loved one (Detering et al. 2010). In addition, recent evidence suggests that early introduction of advance care planning as part of a programme of best palliative care not only leads to a better quality of life in the last months of life but may even prolong life (Temel et al. 2010), while concomitantly reducing costs (Zhang et al. 2009). This is presumably due to the fact that, if they are well informed about alternatives and supported by staff, most patients in an advance care planning process will choose less intensive care, with fewer risks associated with adverse effects of drugs and surgery.

There is no doubt that more evidence is needed, and that the advance care planning process must also be linked to intensive care medicine, to mainstream cardiac, pulmonary and oncology care, and to other important developments, such as evidence-based decision aids (Stacey et al. 2012) and chronic disease self-management programs (e.g. http://patienteducation.stanford.edu/programs/cdsmp.html). However, the diagnosis of the failure of the living will pronounced 10 years ago can now certainly be considered a false positive.

References

Back, Anthony M., Robert M. Arnold, Walter F. Baile, Kelly A. Fryer-Edwards, Stewart A. Alexander, Gwyn E. Barley, Ted A. Gooley, and James A. Tulsky. 2007. Efficacy of communication skills training for giving bad news and discussing transitions to palliative care. *Archives of Internal Medicine* 167: 453–460.

Baile, Walter F., Renato Lenzi, Patricia A. Parker, Robert Buckman, Lorenzo Cohen, R. Lenzi, P.A. Parker, R. Buckman, and L. Cohen. 2002. Oncologists' attitudes toward and practices in giving bad news: An exploratory study. *Journal of Clinical Oncology* 20(8): 2189–2196.

Briggs, Linda. 2003. Shifting the focus of advance care planning: Using an in-depth interview to build and strengthen relationships. *Innovations in End-of Life Care* 5(2): 1–16. www.edc.org/lastacts.

Clayton, J.M., K.M. Hancock, P.N. Butow, M.H. Tattersall, D.C. Currow, Australian and new Zealand Expert Advisory Group, J. Adler, S. Aranda, K. Auret, F. Boyle, A. Britton, R. Chye, K. Clark, P. Davidson, J.M. Davis, A. Girgis, S. Graham, J. Hardy, K. Introna, J. Kearsley, I. Kerridge, L. Kristjanson, P. Martin, A. McBride, A. Meller, G. Mitchell, A. Moore, B. Noble, I. Olver, S. Parker, M. Peters, P. Saul, C. Stewart, L. Swinburne, B. Tobin, K. Tuckwell, P. Yates, Australasian Society of HIV Medicine, Australian and New Zealand Society of Palliative Medicine, Australasian Chapter of Palliative Medicine, Royal Australasian College of Physicians, Australian College of Rural and Remote Medicine, Australian General Practice Network, Australian Society of Geriatric Medicine, Cancer Voices Australia, Cardiac Society of Australia and New Zealand, Clinical Oncological Soceity of Australia, Motor Neurone Disease Association of Australia, Palliative Care Australia, Palliative Care Nurses Australia, Royal Australian College of General Practitioners, Royal College of Nursing, Australia; Thoracic Society of Australia and New Zealand. 2007. Clinical practice guidelines for communicating prognosis and end-of-life issues with adults in the advanced stages of a life-limiting illness, and their caregivers. Medical Journal of Australia 186(12 Suppl):S77, S79, S83–108.

Detering, Karin M., Andrew D. Hancock, Michael C. Reade, and William Silvester. 2010. The impact of advance care planning on end of life care in elderly patients: Randomised controlled trial. *British Medical Journal* 340: c1345.

Fagerlin, Angela, and Carl E. Schneider. 2004. Enough. The failure of the living will. *The Hastings Center Report* 34(2): 30–42.

Fletcher, John C., and Robert Boyle (eds.). 1997. *Introduction to clinical ethics*, 2nd ed. Hagerstown: University Publishing Group.

Hagerty, Rebecca G., Phyllis N. Butow, Peter M. Ellis, Elizabeth A. Lobb, Susan C. Pendlebury, Natasha Leighl, Craig MacLeod, and Martin H.N. Tattersall. 2005. Communicating with realism and hope: Incurable cancer patients' views on the disclosure of prognosis. *Journal of Clinical Oncology* 23(6): 1278–1288.

Hammes, Bernard J., Brenda L. Rooney, and Jakob D. Gundrum. 2010. A comparative, retrospective, observational study of the prevalence, availability, and specificity of advance care plans in a county that implemented an advance care planning microsystem. *Journal of the American Geriatrics Society* 58(7): 1249–1255.

In der Schmitten, Jürgen, Sonja Rothärmel, Christine Mellert, Stephan Rixen, Bernard J. Hammes, Linda Briggs, Karl Wegscheider, and Georg Marckmann. 2011. A complex regional intervention to implement advance care planning in one town's nursing homes: Protocol of a controlled inter-regional study. *BMC Health Services Research* 11: 14.

Nous Foundation. 2009. ACP videos for advance care planning. http://www.acpdecisions.org/videos/. Last accessed 13 Aug 2013.

Rid, Annette, and David Wendler. 2010. Can we improve treatment decision making for incapacitated patients? *The Hastings Center Report* 40(5): 36–45.

Schubart, Jane R., Benjamin H. Levi, Fabian Camacho, Megan Whitehead, Elana Farace, and Michael J. Green. 2012. Reliability of an interactive computer program for advanced care planning. *Journal of Palliative Medicine* 15(6): 637–642.

Shalowitz, David I., Elizabeth Garrett-Mayer, and David Wendler. 2006. The accuracy of surrogate decision makers. A systematic review. *Archives of Internal Medicine* 166: 493–497.

Silvester, William. 2012. Respecting patient choices: Scaling care planning to a whole country. In *Having your own say. Getting the right care when it matters most*, ed. Bernard J. Hammes, 57–70. Washington, DC: CHT Press.

Stacey, Dawn, Carol L. Bennett, Michael J. Barry, Nananda F. Col, Karen B. Eden, Margaret Holmes-Rovner, Hillary Llewellyn-Thomas, Anne Lyddiatt, France Légaré, and Richard Thomson. 2012. Decision aids for people facing health treatment or screening decisions. *Cochrane Library*. doi:10.1002/14651858.CD001431.pub3.

Stiefel, F., J. Barth, J. Bensing, L. Fallowfield, L. Jost, D. Razavi, A. Kiss, and participants. 2010. Communication skills training in oncology: A position paper based on a consensus meeting among European experts in 2009. *Annals of Oncology* 21(2): 204–207.

Temel, Jennifer S., Joseph A. Greer, Alona Muzikansky, Emily R. Gallagher, Sonal Admane, Vicky A. Jackson, Constance M. Dahlin, Craig D. Blinderman, Juliet Jacobson, William F. Pirl, J. Andrew Billings, and Thomas J. Lynch. 2010. Early palliative care for patients with metastatic non small cell lung cancer. *The New England Journal of Medicine* 363: 733–742.

The SUPPORT Principal Investigators. 1995. A controlled trial to improve care for seriously ill hospitalized persons. The study to understand prognoses and preferences for outcomes and risks of treatments (SUPPORT). *Journal of the American Medical Association* 274(20): 1591–1598.

Volandes, Angelo E., Michael K. Paasche-Orlow, Michael J. Barry, Muriel R. Gillick, Kenneth L. Minaker, Yuchiao Chang, Francis E. Cook, Elmer D. Abbo, Areej El-Jawahri, and Susan L. Mitchell. 2009. Video decision support tool for advance care planning in dementia: Randomised controlled trial. *British Medical Journal* 338: b2159.

Weiner, Joseph S., and Steven A. Cole. 2004. ACare: A communication training program for shared decision making along a life-limiting illness. *Palliative & Supportive Care* 2: 231–241.

Wilson, Kent S., and Sue A. Schettle. 2012. Honoring choices Minnesota: A metropolitan program underway. In *Having your own say. Getting the right care when it matters most*, ed. Bernard J. Hammes, 41–56. Washington, DC: CHT Press.

Zhang, Baohui, Alexi A. Wright, Haiden A. Huskamp, Matthew E. Nilsonn, Matthew L. Maciejewski, Craig C. Earle, Susan D. Block, Paul L. Maciejewski, and Holly G. Prigerson. 2009. Health care costs in the last week of life. Associations with end of life conversations. *Archives of Internal Medicine* 169(5): 480–488.

Part V
Conclusions

Chapter 14
Concluding Remarks

Nikola Biller-Andorno, Susanne Brauer, and Peter Lack

Although advance directives have been widely discussed since the 1980s, their ethical basis still remains a matter of heated debate: What makes an advance directive valid, placing others under a moral obligation to follow its instructions? Where should we set ethical boundaries for the scope and binding force of advance directives? What effects do advance directives have on relations with family, loved ones and professionals, and are these effects desirable from a moral point of view? What ethical opportunities and risks are associated with advance directives, given their prerequisites, limitations and effects? No definitive, or even satisfactory, answers have been given to these essential questions. But, especially in view of the increasing prevalence of advance directives in Europe, these questions need to be resolved if advance directives are to be justified as an ethically compelling tool for realizing patient self-determination—a tool worthy of social, political and medical support.

The present volume seeks to contribute to the ongoing debate by integrating fundamental ethical issues with practical matters concerning the implementation of advance directives. The authors highlight cultural, national and professional differences not just in laws and regulations, but also in how advance directives are understood by healthcare professionals and by patients. These views do not necessarily reflect the ways in which advance directives are actually implemented in clinical practice. Revealing such differences and even identifying inconsistencies between conceptions, legal regulations and clinical practice can set the stage for one of the future challenges—tackling the question of whether it is (culturally and politically) practicable and (ethically) required to try and reach a more substantial agreement on advance directives beyond the minimal consensus formulated in

N. Biller-Andorno • S. Brauer (✉)
Institute of Biomedical Ethics, University of Zurich, Zurich, Switzerland
e-mail: nikola.biller@ibme.ch; brauer@ethik.uzh.ch

P. Lack
Department of Moral Theology and Ethics, University of Fribourg, Basel, Switzerland
e-mail: mail@peterlack.ch

P. Lack et al. (eds.), *Advance Directives*, International Library of Ethics, Law, and the New Medicine 54, DOI 10.1007/978-94-007-7377-6_14,
© Springer Science+Business Media Dordrecht 2014

Article 9 of the 1997 Convention on Human Rights and Biomedicine: "The previously expressed wishes relating to a medical intervention by a patient who is not, at the time of the intervention, in a state to express his or her wishes shall be taken into account."

Whatever the legal status of advance directives may be, the authors are convinced that advance directives are ethically valuable because they give a voice to patients at a time when their decision-making capacity has been lost. Granting such a voice is essential, given that the process of dying can be prolonged as a result of advanced medical technologies. In order to provide good end-of-life care and a satisfactory "quality of dying", individual preferences have to be taken into account. Advance directives also enable patients to express their wishes in non-terminal (e.g. psychiatric) conditions. In such cases, benevolence may involve respecting patients' autonomy by acting in accordance with their individual preferences and wishes, especially regarding refusal of treatment.

As demonstrated in several chapters of this volume, advance directives have a significant impact on the patient-healthcare professional relationship and can foster a more patient-centred medicine. This impact is independent of the number of patients who possess an advance directive. Even if the historical advance directive movement has missed its goal—since only a minority of patients make use of advance directives—these documents have been successful in shifting the focus from the healthcare professional to the patient when medical decisions are made.

Regarding advance directives as a way of promoting patient autonomy and well-being—and accepting such an endeavour as a political goal—is compatible with both individually oriented and family-/community-oriented cultures. In embracing a relational conception of autonomy, advance directives are not incompatible with placing a high value on relations with family members or other close persons. This is particularly the case when advance directives are used to designate a proxy for future medical decisions. In fact, a surrogate decision-maker is appointed in the majority of advance directives.

Since advance directives give instructions for future situations of decision-making incapacity, patients are essentially dependent on healthcare providers to act on their wishes. However, advance directives are not always clearly formulated and have to be interpreted in the light of the specific medical situation. From the perspective of the third parties, a significant limitation of advance directives is the inherent uncertainty as to whether the advance directive is actually applicable in the given situation. As research shows, third parties often have difficulty in correctly judging patients' wishes and preferences. Their perspectives always remain external to the patient, while the advance directive at least expresses the patient's perspective.

In order to ease the potential tension between the perspectives of the patient and healthcare providers and to facilitate anticipation of future medical situations, it would be helpful to place advance directives in the broader context of advance care planning, in which communication with healthcare professionals is supported and decision-making is shifted towards a patient-oriented process. Advance directives themselves can then function as an accepted tool for communication between the parties concerned both at the drafting and the implementation stage. This strengthens the validity and binding force of advance directives from the perspective of third parties—which is crucial, as they are responsible for implementation. It has also been shown empirically that advance care planning serves to improve end-of-life care.

The use of advance directives also involves certain risks. Some patients fear that if they write an advance directive, they will not be treated by the same standards as those who do not—i.e. they might experience a limitation of treatment, not reflecting their previously expressed wishes. Some healthcare institutions might indeed be tempted to require a care-limiting advance directive before admitting patients. Such fears could be fuelled by discussions about the impact of advance directives on cutting healthcare costs. Although reducing expenses is economically necessary even in affluent countries—especially in view of the high costs of end-of-life treatment—it would clearly be an abuse if advance directives were employed against patients' wishes for financial reasons.

In order to make advance directives an effective tool for self-determination and to prevent misuse, two issues need to be resolved. Firstly, as regards the binding force of advance directives, it is essential that there should be legal safeguards in case the patient's representative or medical staff do not act in accordance with the patient's wishes. Moral appeals to respect the patient's wishes are not sufficient to guarantee patient autonomy in clinical practice. In this respect, further elaboration of the European Convention on Human Rights and Biomedicine is to be recommended; amendments are also required where no provision is made in national legislation for a patient, proxy or loved ones to appeal if they are convinced that the patient's wishes are not being respected. The availability of legal recourse would in itself have a positive effect on clinical practice, giving healthcare staff such as nurses the courage to express their moral concerns when care-related conflicts arise.

Secondly, the decision to declare a patient incompetent is crucial for the patient's involvement in the medical decision-making process. The declaration of incompetence is the trigger for implementation of an advance directive. However, standard tools for assessing decision-making capacity need to be further refined and, especially, harmonized and consistently implemented in clinical practice. The elaboration and use of such tools must also be guided by ethical considerations.

Improving our understanding of when advance directives should take effect and ensuring that they are appropriately managed would be an important step towards a healthcare system in which patient-oriented outcomes are taken seriously, even at times when it is difficult to ascertain what is in the patient's interest.

Information on Editors and Authors

Jacqueline M. Atkinson, Ph.D., C.Psychol., is an Honorary Senior Research Fellow in Public Health at the University of Glasgow (UK). Formerly Professor of Mental Health Policy and Chair of the University of Glasgow Research Ethics Committee. Served as an adviser to the Scottish Parliament for the passage of the Mental Health (Care and Treatment) Act through Parliament. Current research interests: mental health law, advance directives in mental health, mental health policy.

Mark P. Aulisio, Ph.D., is Professor of Bioethics at Case Western Reserve University and Director of the Center for Biomedical Ethics, MetroHealth System, Cleveland, Ohio (US). Degree in applied philosophy. Helped to lead three American Society for Bioethics and Humanities projects in the field of ethics consultation training and education and received the ASBH Distinguished Service Award (2009). Current research interests: clinical ethics policy and practice, intersection of bioethics and political philosophy.

Sohaila Bastami, Dr. med., is a Ph.D. student at the Institute of Biomedical Ethics, University of Zurich (Switzerland). Degree in medicine. Former Fellow at the Division of Medical Ethics, Harvard Medical School (2011–2012). Current research interests: clinical ethics (especially transplantation ethics), research ethics and the interface of evidence-based medicine and ethics.

Nikola Biller-Andorno, Professor Dr. med. Dr. phil., is the Director of the Institute of Biomedical Ethics, University of Zurich (Switzerland). Degrees in medicine and philosophy. Past President of the International Association of Bioethics and member of the Central Ethics Committee of the Swiss Academy of Medical Sciences. Harkness/Careum Fellow and Visiting Professor at Harvard University (2012–2013). Current research interests: interface of ethics, economics, health policy and healthcare practice.

Susanne Brauer, Ph.D., is Director of Studies for bioethics, medicine and life sciences at the Paulus-Akademie Zurich (Switzerland). Degree in philosophy and German literature and linguistics. Affiliated Research Fellow at the Institute of Biomedical Ethics, University of Zurich. Scientific collaborator of the Swiss National Advisory Commission on Biomedical Ethics (2008–2012). Current

P. Lack et al. (eds.), *Advance Directives*, International Library of Ethics, Law, and the New Medicine 54, DOI 10.1007/978-94-007-7377-6,
© Springer Science+Business Media Dordrecht 2014

research interests: medical humanities, intersections between politics, healthcare and life sciences.

Bernice Elger, Professor Dr. med., MA theol., is Director of the Institute of Biomedical Ethics, University of Basel (Switzerland). Member of the Expert Commission for Human Genetic Testing (GUMEK) and the Federal Commission on Basic Health Care Services and Fundamental Principles (ELGK). Current research interests: ethical issues relating to cutting-edge biotechnology, biobanking and prison medicine.

Ruth Horn, Ph.D., is an Ethics and Society Wellcome Trust Fellow at the Ethox Centre, University of Oxford (UK). Degrees in sociology. Past Marie Curie Fellow (EC) at the Centre for Ethics in Medicine, University of Bristol (UK). Current research interests: sociology of ethics, end-of-life decision-making, advance directives, euthanasia, terminal sedation.

Ralf J. Jox, M.D., Ph.D., is Assistant Professor of Medical Ethics at the Institute of Ethics, History and Theory of Medicine, Ludwig Maximilian University of Munich (Germany). Studied medicine at the University of Munich and Harvard Medical School, and philosophy and ethics at the Munich School of Philosophy, the University of Basel (Switzerland) and King's College London (UK). Caroline Miles Visiting Scholar at the Ethox Centre, University of Oxford (UK) in 2012. Current research interests: patient autonomy, treatment decision-making, advance care planning, neuroethics, vulnerable patients (especially dementia and disorders of consciousness) and clinical ethics consultation.

Tanja Krones, P.D., Dr. med. Dipl. Soz., is a clinical ethicist and Director of the Clinical Ethics Committee at Zurich University Hospital (Switzerland). Degrees in medicine, sociology and clinical ethics. Member of the Central Ethics Committee of the Federal Chamber of Physicians and of the German Central Ethics Commission on Stem Cell Research. Head of the German Academy for Ethics in Medicine's Working Group on Reproductive Biomedicine, and of the German Network for Evidence-Based Medicine's Working Group on Ethics and EBM. Current research interests: advance care planning, decision aids, shared decision-making, health technology assessment and transplantation.

Peter Lack, Lic. theol., is a medical ethicist, independent project consultant and scientific collaborator at the University of Fribourg (Switzerland). Chaired the Swiss Academy of Medical Sciences committee responsible for the publication of medical-ethical guidelines on advance directives in 2009. Scientific adviser to the Swiss Red Cross in various projects and services concerning patient self-determination. Lecturer in medical ethics, patient counselling, physician-patient communication and pastoral psychology (clinical pastoral education).

Margot Michel, Dr. iur., is a postdoctoral research associate and lecturer in medical law and family law at the Institute of Law, University of Zurich (Switzerland). Serves as legal adviser to the Child and Adult Protection Authority of Canton Zug (Switzerland). Current research interests: medical law, family law, adult and child protection law, property law and animal law.

Christine Mitchell, RN, MS, MTS, is Associate Director of Clinical Ethics at the Division of Medical Ethics, Harvard Medical School, and Director of the Office of

Ethics at Boston Children's Hospital. Past President of the American Society of Law, Medicine and Ethics, a founding member of the Society for Bioethics Consultation, and a member of the American Society for Bioethics and Humanities' Clinical Ethics Consultation Affairs Committee (2010–2013). Current research interests: paediatric ethics and ethics consultation activities.

Settimio Monteverde, MME, MAE, Lic. theol., is a lecturer in nursing ethics at the Bern University of Applied Sciences (Switzerland). Degrees in medical education, applied ethics and theology, diploma as a registered nurse and nurse anaesthetist. Member of the Central Ethics Committee of the Swiss Academy of Medical Sciences. Ph.D. student (Careum Fellow) at the Institute of Biomedical Ethics, University of Zurich (Switzerland). Current research interests: healthcare ethics education, nursing ethics, ethics of palliative care.

Robert S. Olick, JD, Ph.D., is Associate Professor at the Center for Bioethics and Humanities, SUNY Upstate Medical University, New York (US), where he teaches bioethics and health law for medical and allied health professional students. Previously served as Executive Director of the New Jersey Bioethics Commission. Current research interests: end-of-life decisions, genetic privacy and discrimination, ethics consultation, physician-patient-family relationship, health policy.

Jacqueline Reilly, who has a degree in social sciences, is a University Teacher in Public Health at the University of Glasgow School of Medicine. Previously a researcher in public health and social sciences, also at Glasgow. Current research interests: public health policy, risk behaviours and mental health law.

Alfred Simon, Dr. Phil., is Executive Director of the German Academy for Ethics in Medicine, Göttingen (Germany). Degree in philosophy. Chair of the Health Care Ethics Committee of Göttingen University Hospital. Member of the Ethical and Legal Advisory Board of the German Medical Association. Current research interests: clinical ethics, end-of-life decision-making, patient autonomy, advance directives, ethics in psychiatry.

Ruud ter Meulen, Ph.D., a psychologist and ethicist, is Professor of Ethics in Medicine and Director of the Centre for Ethics in Medicine at the University of Bristol (UK). Previously Professor of Philosophy and Medical Ethics and Director of the Institute for Bioethics at the University of Maastricht (the Netherlands). Current research interests: solidarity and justice in healthcare, healthcare reform and health policy, evidence-based medicine, long-term care, research ethics and research ethics committees.

Marie-Jo Thiel, Professor Dr. med., Dr. theol., is Director of the European Centre for the Study and Teaching of Ethics, University of Strasbourg (France). Degrees in medicine, theological ethics and European health policy. Member of the European Group on Ethics in Science and New Technologies. Director of the "Bioethics and Religion" Research Group. Member of the Ethics Committee of the Faculty of Medicine in Strasbourg and of the Ethics Committee of the French Society for Thoracic and Cardiovascular Surgery. Current research interests: interface of ethics, healthcare, bioethics, theological ethics, disability and ageing.

Manuel Trachsel, Dr. Phil., B. Med., is a postdoctoral research associate at the Institute of Biomedical Ethics, University of Zurich (Switzerland). Studied psychology, medicine and philosophy/ethics at the University of Bern (Switzerland). Current research interests: clinical psychology, psychotherapy research, clinical decision-making and end-of-life issues.

Jochen Vollmann, Professor Dr. med., Dr. phil., a psychiatrist and psychotherapist, is Director of the Institute for Medical Ethics and History of Medicine and President of the Centre for Medical Ethics at Ruhr University Bochum (Germany). Current research interests: informed consent and capacity assessment, ethics and psychiatry, advance directives, end-of-life decision-making, personalized medicine, clinical ethics committees and clinical ethics consultation.

Index

CPSIA information can be obtained at www.ICGtesting.com
Printed in the USA
LVOW10*2137071113

360485LV00009B/74/P